AF531332

CHEMICAL BASIS OF INHERITANCE

ENCYCLOPAEDIA OF GENETICS

Vol. III

CHEMICAL BASIS OF INHERITANCE

By

Dr. Arvind N. Shukla

School of Studies of Zoology & Biotechnology

Vikram University

Ujjain

DISCOVERY PUBLISHING HOUSE PVT. LTD.

NEW DELHI-110 002

First Published-2008

ISBN 978-81-8356-304-8 (Set)

Published by:

DISCOVERY PUBLISHING HOUSE PVT. LTD.

4831/24, Ansari Road, Prahlad Street,
Darya Ganj, New Delhi-110002 (India)
Phone: 23279245 • Fax: 91-11-23253475
E-mail: dphbooks@rediffmail.com
dphtemp@indiatimes.com

Printed by : Sachin Printers, Delhi

Dedicated to

Prof. Ram Rajesh Mishra

Hon'ble Vice Chancellor

Vikram University, Ujjain [M.P.]

Preface

The present title *"Encyclopaedia of Genetics"* is an exciting, and dynamic branch of science and offers the finest approach to teaching genetics through the integration of the molecular and chemical subdisciplines. It prepares the students to learn to formulate genetic hypothesis and apply critical thinking skill necessary for problem solving, while also gaining a sense of the social and historical context in which genetics has developed. This text also has a completely novel way to illustrate the one or two experiments in each chapter that are rigorously examined according to the scientific method. It starts with the premise that the syllabus for a university course in genetics should reflect the major research issues of the new millennium rather than those topics that were in vogue during the last decades of nineteenth century.

The text covers both the basic and practical aspects of Genetics. Fundamental knowledge is developed within the context of applied relevance. Principles are supplemented with examples. This is done to maintain student interest, which is essential for learning any subject.

In the preparation of this book large number of books and research papers have been consulted. So no authenticity is claimed.

The author expresses his gratitude to Mr. Wasan and staff of M/s Discovery Publishing House for their whole hearted co-operation in the publication of this book.

The author tried hard to be accurate and upto date in statement and realises the impossibility of completely avoiding errors therefore, the author will greatly appreciate having his attention called to any questionable statement.

Author

CONTENTS

1

Chemistry of Gene

In 1953, James Watson and Francis Crick published a two-page paper in the journal *Nature* entitled "Molecular Structure of Nucleic Acids: A Structure for Deoxyribose Nucleic Acid." It began as follows:" We wish to suggest a structure for the salt of deoxyribose nucleic acid (D.N.A.). This structure has novel features which are of considerable biological interest." This paper, which first put forth the correct model of DNA structure, is a milestone in the modern era of molecular genetics, compared by some to the work of Mendel and Darwin. (Watson, Crick, and X-ray crystallographer Maurice Wilkins won Nobel Prizes for this work; Rosalind Franklin, also an X-ray crystallographer, was acknowledged, posthumously, to have played a major role in the discovery of the structure of DNA.) Once the structure of the genetic material had been determined, an understanding of its method of replication and its functioning quickly followed.

In Search of the Genetic Material

Required Properties of a Genetic Material

We begin with a look at the properties that a genetic material must have and review the evidence that nucleic acids make up the genetic material. To comprise the genes, DNA must carry the information to control the synthesis of the enzymes and proteins within a cell or organism; self-replicate with high fidelity, yet show a low level of mutation; and be located in the chromosomes.

Control of the proteins

The growth, development, and functioning of a cell are controlled by the proteins within it, primarily its enzymes. Thus, the nature of a cell's phenotype is controlled by the protein synthesis within that cell. The genetic material must therefore determine the need for and effective amounts of the enzymes in a cell. For example, given inorganic salts and glucose, an *E. coli* cell can synthesize, through its enzyme-controlled biochemical pathways, all of the compounds it needs for growth, survival, and reproduction. In contrast, a mammalian red blood cell primarily produces hemoglobin.

At this point we need to review some basic information regarding enzymes. An enzyme is a protein that acts as a catalyst for a specific metabolic process without itself being markedly altered by the reaction. Most reactions that enzymes catalyze could occur anyway, but only under conditions too extreme to take place within living systems. For example, many oxidations occur naturally at high temperatures. Enzymes allow these reactions to occur within the cell by lowering the *free energy of activation* ($\Delta G^{\ddagger}$) of a particular reaction. In other words, an enzyme allows a reaction to take place without needing the boost in energy that heat usually supplies.

Most metabolic processes, such as the biosynthesis or degradation of molecules, occur in pathways, with enzyme facilitating each step in the pathway. Each reaction product in the pathway is altered by an enzyme that converts it to the next product. The enzyme threonine dehydratase, for example, converts threonine into α-ketobutyric acid. Enzymes are composed of folded polymers of amino acids. The average protein is three hundred to five hundred amino acids long; only twenty naturally occurring amino acids are used in constructing these proteins. The sequence of amino acids determines the final structure of an

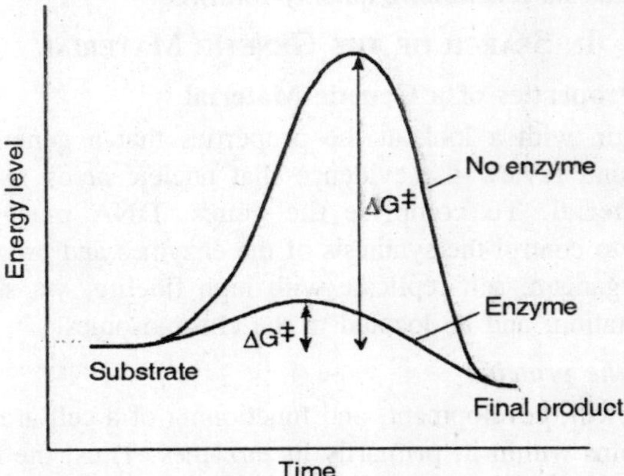

Fig. 1.1. An enzyme lowers the free energy of activation ($\Delta G^{\ddagger}$) for a particular reaction.

enzyme. The genetic material determines the sequence of the amino acids.

The three-dimensional structure of enzymes permits them to perform their function. An enzyme combines with its substrate or substrates (the molecules it works on) at a part of the enzyme called the *active site*. The substrates "fit" into the active site, which has a shape that allows only the specific substrates to enter. This view of the way an enzyme interacts with its substrates is called the *lock-and-key model* of enzyme functioning. When the substrates are in their proper position in the active site of the enzyme, the particular reaction that the enzyme catalyzes takes place. The reaction products then separate from the enzyme and leave it free to repeat the process. Enzymes can work at phenomenal speeds. Some can catalyze as many as a million reactions per minute.

Not all of the cell's proteins function as catalysts. Some are structural proteins, such as keratin, the main component of hair. Other proteins are regulatory—they control the rate at which other enzymes work. Still others are involved in different functions; albumins, for example, help regulate the osmotic pressure of blood.

Replication

The genetic material must be capable of precisely directing its own replication so that every daughter cell receives an exact copy. Some *mutability*, or the ability to change, is also required, because we know that the genetic material has changed, or evolved, over the history of life on earth. In their 1953 paper, Watson and Crick had already worked out the replication process based on the structure of DNA. The fidelity of the replication process is so great that the error rate is only about one in a billion.

Location

It has been known since the turn of the century that genes, the discrete functional units of genetic material, are located in chromosomes within the nuclei of eukaryotic cells: the way chromosomes behave during the cellular division stages of mitosis and meiosis mimics the behavior of genes. Thus, the genetic material in eukaryotes must be a part of the chromosomes.

For a long time, proteins were considered the most probable genetic material because they have the necessary molecular complexity. The twenty naturally occurring amino acids can be combined in an almost unlimited variety, creating thousands and thousands of different proteins.

The first proof that the genetic material is deoxyribonucleic acid (DNA) came in 1944 from Oswald Avery and his colleagues. The Watson and Crick model in 1953 ended a period when many thought DNA was the genetic material, but its structure was unknown.

Evidence for DNA as the Genetic Material

Transformation

In 1928, F. Griffith reported that heat-killed bacteria of one type could "transform" living bacteria of a different type. Griffith demonstrated this transformation using two strains of the bacterium *Streptococcus pneumoniae*. One strain (S) produced smooth colonies on media in a petri plate because the cells had polysaccharide capsules. It caused a fatal bacteremia (bacterial infection) in mice. Another strain (R), which lacked polysaccharide capsules, produced rough colonies on petri plates; it did not have a pathological effect on mice.

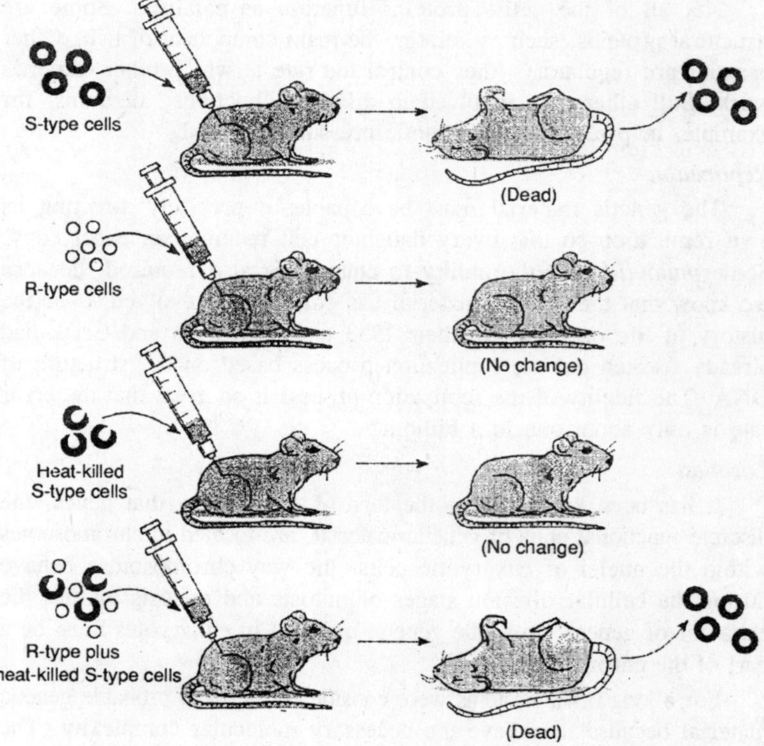

Fig. 1.2. Griffith's experiment with Streptococcus.

Bacteria of the rough strain are engulfed by the mice's white blood cells; bacteria of the virulent smooth strain survive because their polysaccharide coating protects them. Griffith found that neither heat-killed S-type nor live R-type cells, by themselves, caused bacteremia in mice. However, if he injected a mixture of live R-type and heat-killed S-type cells into mice, the mice developed a bacteremia identical to that caused by living S-type cells. Thus, something in the heat-killed S cells transformed the R-type bacteria into S-type cells.

In 1944, Oswald Avery and two of his associates, C. MacLeod and M. McCarty, reported the nature of the transforming substance. Avery and his colleagues did their work *in vitro* (literally, in glass), using colony morphology on culture media rather than bacteremia in mice as evidence of transformation. They ruled out proteins, carbohydrates, and lipids by their extraction procedure, by the chemical analysis of the transforming material, and by demonstrating that the only enzymes that destroyed the transforming ability were enzymes that destroyed DNA. This study provided the first experimental evidence that DNA was the genetic material: DNA transformed R-type bacteria into S-type bacteria.

Phage labeling

Valuable information about the nature of the genetic material has also come from viruses. Of particular value are studies of bacterial viruses—the bacteriophages, or phages. Since phages consist only of nucleic acid surrounded by protein, they lend themselves nicely to the determination of whether the protein or the nucleic acid is the genetic material.

A. D. Hershey and M. Chase published, in 1952, the results of research that supported the notion that DNA is the genetic material and, in the process, helped to explain the nature of the viral infection process. Since all nucleic acids contain phosphorus, whereas proteins do not, and since most proteins contain sulfur (in the amino acids cysteine and methionine), whereas nucleic acids do not, Hershey and Chase designed an experiment using radioactive isotopes of sulfur and phosphorus to keep separate track of the viral proteins and nucleic acids during the infection process. They used the T2 bacteriophage and the bacterium *Escherichia coli*. The phages were labeled by having them infect bacteria growing in culture medium containing the radioactive isotopes ^{35}S or ^{32}P. Hershey and Chase then proceeded to identify the material injected into the cell by phages attached to the bacterial wall.

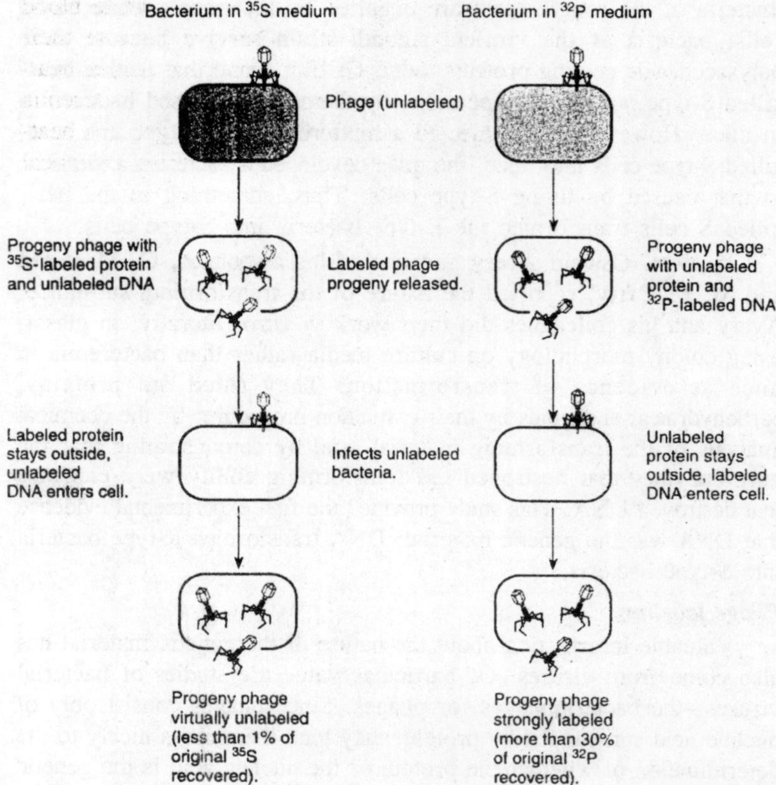

Fig. 1.3. The Hershey and Chase experiments using ^{35}S-labeled and ^{32}P-labeled T2 bacteriophages. The nucleic acid lebel (^{32}P) enters the E. coli bacteria during infection; the protein label (^{35}S) does not.

When ^{32}P-labeled phages were mixed with unlabeled *E. coli* cells, Hershey and Chase found that the ^{32}P label entered the bacterial cells and that the next generation of phages that burst from the infected cells carried a significant amount of the ^{32}P label. When ^{35}S-labeled phages were mixed with unlabeled *E. coli*, the researchers found that the ^{35}S label stayed outside the bacteria for the most part. Hershey and Chase thus demonstrated that the outer protein coat of a phage does not enter the bacterium it infects, whereas the phage's inner material, consisting of DNA, does enter the bacterial cell. Since the DNA is responsible for the production of the new phages during the infection process, the DNA, not the protein, must be the genetic material.

RNA as genetic material

In some viruses, RNA (ribonucleic acid) is the genetic material. The tobacco mosaic virus that infects tobacco plants consists only of RNA and protein. The single, long RNA molecule is packaged within a rodlike structure formed by over two thousand copies of a single protein. No DNA is present in tobacco mosaic virus particles. In

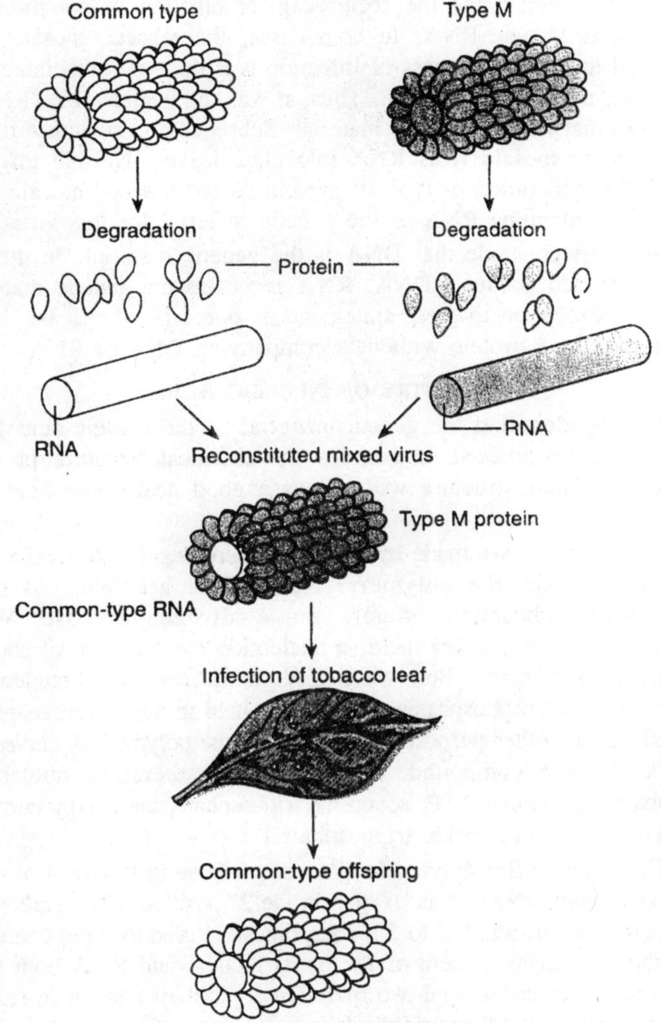

Fig. 1.4. Reconstitution experiment of Fraenkel-Conrat and Singer.

1955, H. Fraenkel-Conrat and R. Williams showed that a virus can be separated, in vitro, into its component parts and reconstituted as a viable virus. This finding led Fraenkel-Conrat and B. Singer to reconstitute tobacco mosaic virus with parts from different strains. For example, they combined the RNA from the common tobacco mosaic virus with the protein from the masked (M) strain of tobacco mosaic virus. They then made the reciprocal combination of common-type protein and M-type RNA. In both cases, the tobacco mosaic virus produced during the process of infection was the type associated with the RNA, not with the protein. Thus, it was the nucleic acid (RNA in this case) that was the genetic material. Subsequently, scientists rubbed pure tobacco mosaic virus RNA into plant leaves. Normal infection and a new generation of typical, protein-coated tobacco mosaic virus resulted, confirming RNA as the genetic material for this virus.

We thus conclude that DNA is the genetic material. In the few viruses that do not have DNA, RNA serves as the genetic material. The only exception to these statements is one type of disease that is transmitted by a protein without accompanying DNA or RNA.

Chemistry of Nucleic Acids

Having identified the genetic material as the nucleic acid DNA (or RNA), we proceed to examine the chemical structure of these molecules. Their structure will tell us a good deal about how they function.

Nucleic acids are made by joining *nucleotides* in a repetitive way into long, chain-like polymers. Nucleotides are made of three components: phosphate, sugar, and a nitrogenous base. When incorporated into a nucleic acid, a nucleotide contains one of each of the three components. But, when free in the cell pool, nucleotides usually occur as triphosphates. The energy held in the extra phosphates is used, among other purposes, to synthesize the polymer. A *nucleoside* is a sugar-base compound. Nucleotides are therefore nucleoside phosphates. (Note that ATP, adenosine triphosphate, the energy currency of the cell, is a nucleoside triphosphate.)

The sugars differ only in the presence (ribose in RNA) or absence (deoxyribose in DNA) of an oxygen in the 2' position. (The carbons of the sugars are numbered 1' to 5'. The primes are used to avoid confusion with the numbering system of the bases.) DNA and RNA both have four bases (two *purines* and two *pyrimidines*) in their nucleotide chains. Both molecules have the purines *adenine* and *guanine* and the pyrimidine *cytosine*. DNA has the pyrimidine *thymine*; RNA has the pyrimidine

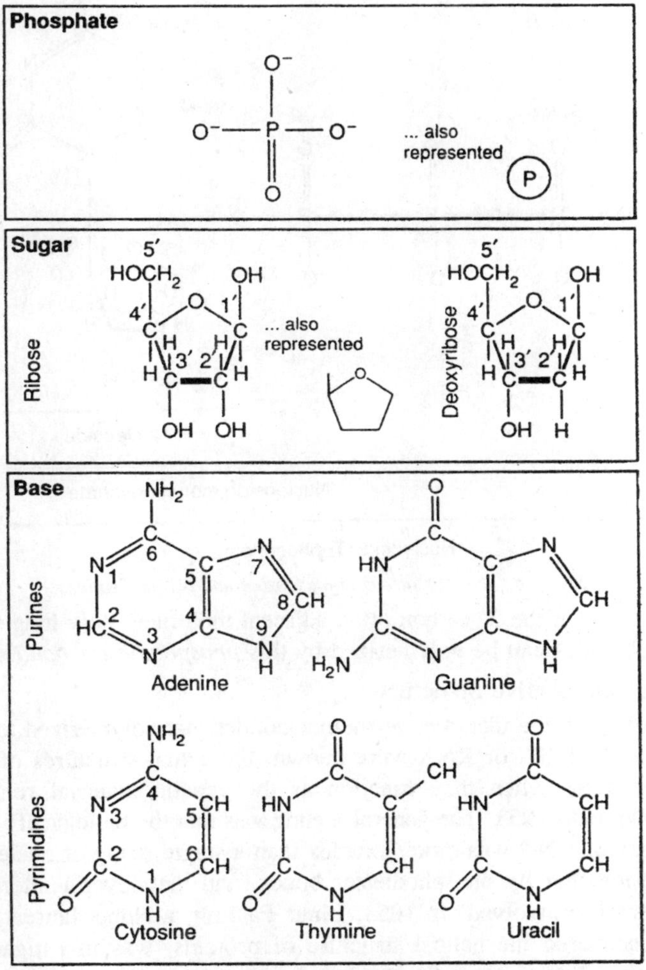
Phosphate
O⁻
O⁻—P—O⁻
O
... also represented
P
Sugar
Ribose
5′
HOCH₂
OH
4′
1′
3′ 2′
OH OH
... also represented
Deoxyribose
OH H
Base
Purines
NH₂
Adenine
Guanine
Pyrimidines
Cytosine
Thymine
Uracil

Fig. 1.5. Components of nucleic acids: phosphate, sugar, and bases.

uracil. Thus, three of the nitrogenous bases are found in both DNA and RNA, whereas thymine is unique to DNA, and uracil is unique to RNA.

A nucleotide is formed in the cell when a base attaches to the 1' carbon of the sugar and a phosphate attaches to the 5' carbon of the same sugar; the nucleotide takes its name from the base. Nucleotides are linked together (*polymerized*) by the formation of a bond between the phosphate at the 5' carbon of one nucleotide and the hydroxyl

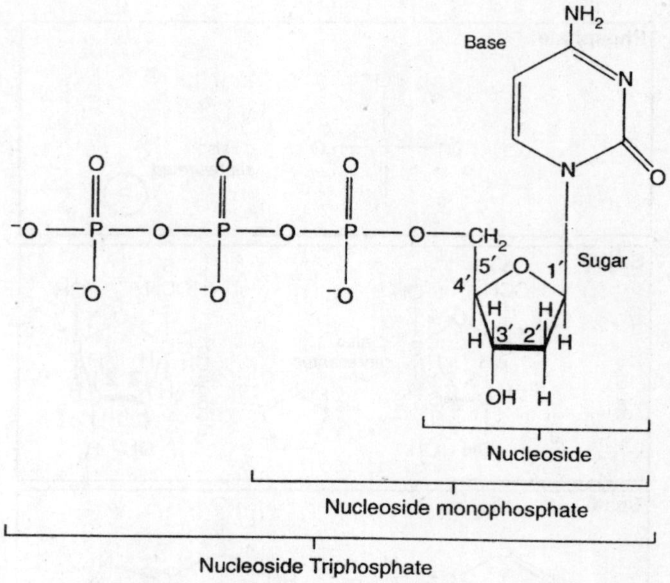

Fig. 1.6. The structure of a nucleoside and two nucleotides.

(OH) group at the 3' carbon of an adjacent molecule. Very long strings of nucleotides can be polymerized by this *phosphodiester bonding*.

Biologically Active Structure

Although the identities of the nucleotides that polymerized to form a strand of DNA or RNA were known, the actual structures of these nucleic acids when they function as the genetic material remained unknown until 1953. The general feeling was that the biologically active structure of DNA was more complex than a single string of nucleotides linked together by phosphodiester bonds, and that several interacting strands were involved. In 1953, Linus Pauling, a Nobel laureate who had discovered the helical structure of proteins, was investigating a three-stranded structure for the genetic material, whereas Watson and Crick had decided that a two-stranded structure was more consistent with available evidence. Three lines of evidence directed Watson and Crick: the chemical nature of the components of DNA, X-ray crystallography, and Chargaff's ratios.

DNA X-ray crystallography

All the time Watson and Crick were studying DNA structure, Maurice Wilkins, Rosalind Franklin, and their colleagues were using *X-ray crystallography* to analyze the structure of DNA. The molecules

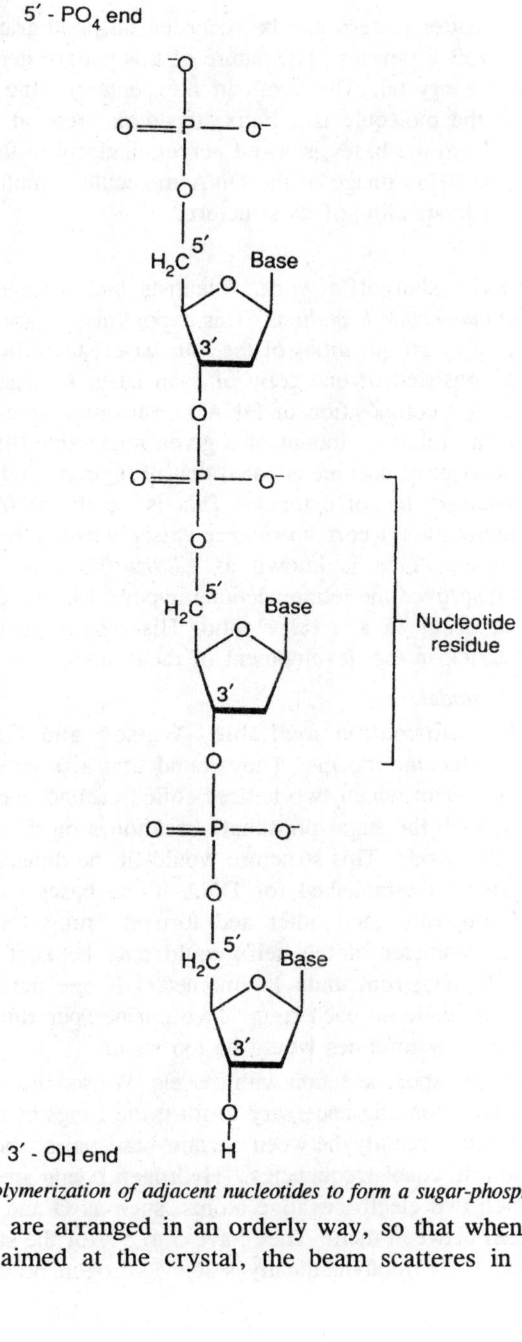

Fig. 1.7. Polymerization of adjacent nucleotides to form a sugar-phosphate strand.

in a crystal are arranged in an orderly way, so that when a beam of X rays is aimed at the crystal, the beam scatteres in an orderly fashion. The scatter pattern can be recorded on photographic film or computer-controlled devices. The nature of this pattern depends on the structure of the crystal. The cross in the center of the photograph indicates that the molecule is a helix; the dark areas at the top and bottom come from the bases, stacked perpendicularly to the main axis of the molecule. This image of the DNA molecule stimulated Watson and Crick's understanding of its structure.

Chargaff's ratios

Until Erwin Chargaff's work, scientists had labored under the erroneous *tetranucleotide hypothesis*. This hypothesis proposed that DNA was made up of equal quantities of the four bases; therefore, a subunit of this DNA consisted of one copy of each base. Chargaff carefully analyzed the base composition of DNA in various species. He found that although the relative amount of a given nucleotide differs among species, the amount of adenine equaled that of thymine and the amount of guanine equaled that of cytosine. That is, in the DNA of all the organisms studied, a 1:1 correspondence exists between the purine and pyrimidine bases. This is known as *Chargaff's rule*. Chargaff's observations disproved the tetranucleotide hypothesis; the four bases of DNA did not occur in a 1:1:1:1 ratio. His results gave insight to Watson and Crick in the development of their model.

Watson-Crick model

With the information available, Watson and Crick began constructing molecular models. They found that a possible structure for DNA was one in which two helices coiled around one another (a *double helix*), with the sugar-phosphate backbones on the outside and the bases on the inside. This structure would fit the dimensions X-ray crystallography had established for DNA if the bases from the two strands were opposite each other and formed "rungs" in a helical "ladder". The diameter of the helix could only be kept constant at about 20 Å (10 angstrom units 1 nanometer) if one purine and one pyrimidine base made up each rung. Two purines per rung would be too big, and two pyrimidines would be too small.

After further experimentation with models, Watson and Crick found that the hydrogen bonding necessary to form the rungs of their helical ladder could occur readily between certain base pairs, the pairs that Chargaff found in equal frequencies. (Hydrogen bonds are very weak bonds in which two electronegative atoms, such as O and N, share a hydrogen atom between them. They have 3 to 5% of the strength of a covalent bond.) Thermodynamically stable hydrogen bonding occurs

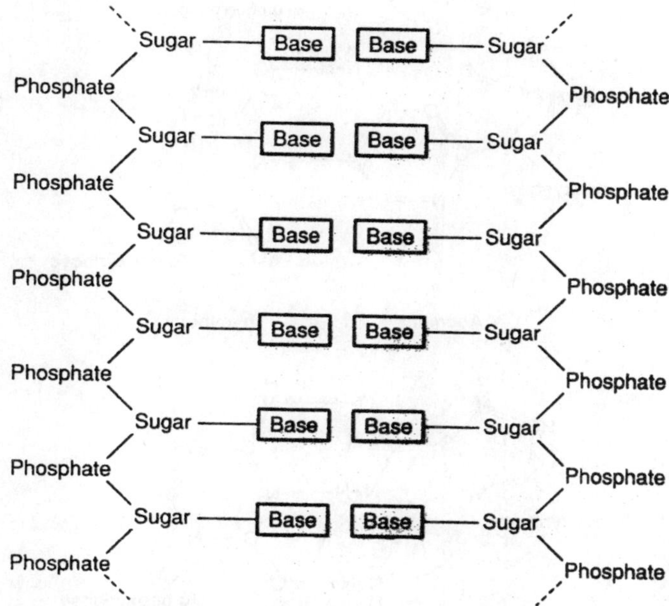

Fig. 1.8. Component parts of double helical structure of DNA.

between thymine and adenine and between cytosine and guanine. The relationship is one of *complementarity*. There are two hydrogen bonds between adenine and thymine and three between cytosine and guanine.

Another point about DNA structure relates to the *polarity* that exists in each strand. That is, one end of a DNA strand has a 5' phosphate and the other end has a 3' hydroxyl group. Watson and Crick found that hydrogen bonding would occur if the polarity of the two strands ran in opposite directions; that is, if the two strands were *antiparallel*.

DNA denaturation

Denaturation studies indicated that the hydrogen bonding in DNA occurs in the way Watson and Crick suggested. Hydrogen bonds, although individually very weak, give structural stability to a molecule in large enough numbers. However, the hydrogen bonds can be broken and the DNA strands separated when the DNA molecule is heated in water. At a certain point, the thermal agitation overcomes the hydrogen bonding, and the molecule becomes *denatured* (or "melts"). It is logical that the more hydrogen bonds DNA contains the higher the temperature needed to denature it. It thus follows that since a G-C (guanine-cytosine)

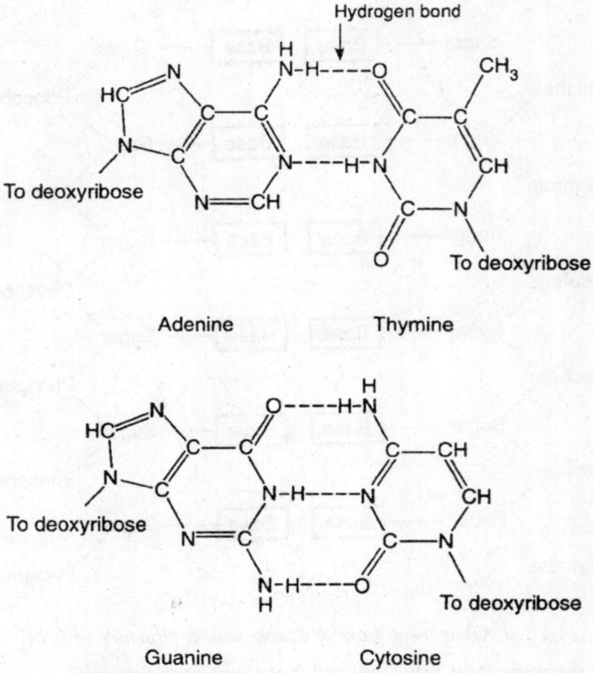

Hydrogen bond

To deoxyribose

To deoxyribose

Adenine Thymine

To deoxyribose

To deoxyribose

Guanine Cytosine

Fig. 1.9. Hydrogen bonding between the nitrogenous bases in DNA.

base pair has three hydrogen bonds to every two in an A-T (adenine-thymine) base pair, the higher the G-C content in a given molecule of DNA, the higher the temperature required to denature it. This relationship exists.

Requirements of Genetic Material

Let us now return briefly to the requirements we have said a genetic material needs to meet: (1) control of protein synthesis, (2) self-replication, and (3) location on the chromosomes in the nucleus (in organisms with nuclei). Does DNA (or when DNA is absent, RNA) meet these requirements?

Control of enzymes

In the next, we examine the details of protein synthesis. We will see that DNA does possess the complexity required to direct protein synthesis. Although complementarity restricts the base opposite a given base in a double helix, there are no restrictions on the sequence of bases on a given strand. Later, we will show that each sequence of three bases in DNA specifies a particular amino acid during protein

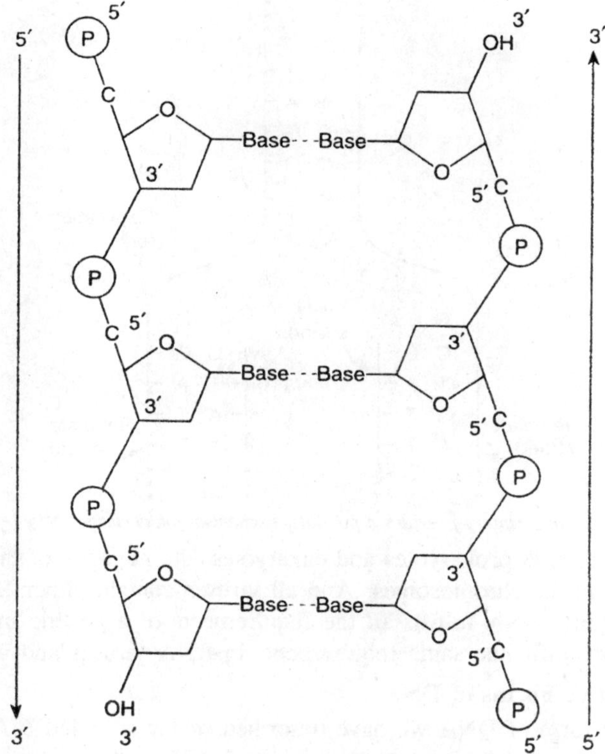

Fig. 1.10. Polarity of the DNA strands.

synthesis. The *genetic code* gives the relationship of DNA bases to the amino acids in proteins.

Replication

Watson and Crick hinted in their 1953 paper how DNA might replicate. Their observation stemmed from the property of complementarity. Since the base sequence on one strand is complementary to the base sequence on the opposite strand, each strand could act as a template for a new double helix if the molecule simply "unzipped," allowing each strand to specify the sequence of bases on a new strand by complementarity. Mutability would occur due to mispairings, other errors in replication, or damage to the DNA.

Location

DNA must reside in the nucleus of eukaryotes, where the genes occur on chromosomes, or in the chromosomes of prokaryotes and

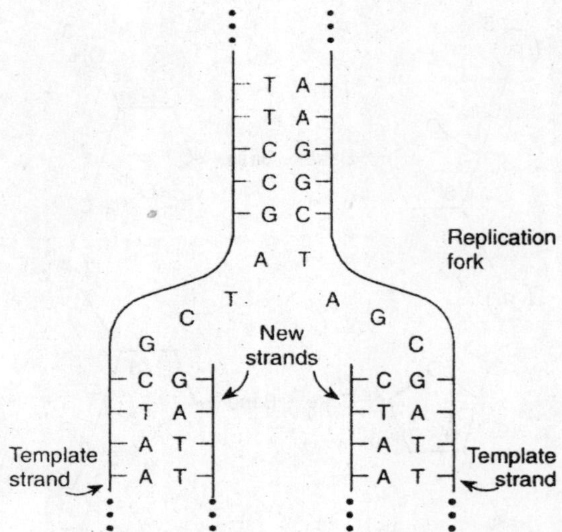

Fig. 1.11. Complementary provides a possible mechanism for accurate DNA replication.

viruses. In both prokaryotes and eukaryotes, the majority of the cell's DNA is in the chromosomes. And all viruses contain either DNA or RNA. Thus, DNA fulfills all the requirements of a genetic material. RNA can fulfill the same requirements in RNA viruses and viroids.

Alternative Forms of DNA

The form of DNA we have described so far is called *B DNA*. It is a right-handed helix: it turns in a clockwise manner when viewed down its axis. The bases are stacked almost exactly perpendicular to the main axis, with about ten base pairs per turn (34 Å). However, DNA can exist in other forms. If the water content increases to about 75%, the A form of DNA (*A DNA*) occurs. In this form, the bases tilt in regard to the axis, and there are more base pairs per turn. However, this and other known forms of DNA are relatively minor variations on the right-handed B form.

In 1979, Alexander Rich and his colleagues at MIT discovered a left-handed helix that they called *Z DNA* because its backbone formed a zigzag structure. Z DNA was found by X-ray crystallographic analysis of very small DNA molecules composed of repeating G-C sequences on one strand with the complementary C-G sequences on the other (alternating purines and pyrimidines). Z DNA looks like B DNA with each base rotated 180 degrees, resulting in a zigzag, left-handed structure. (The original configuration of the bases is referred to as the

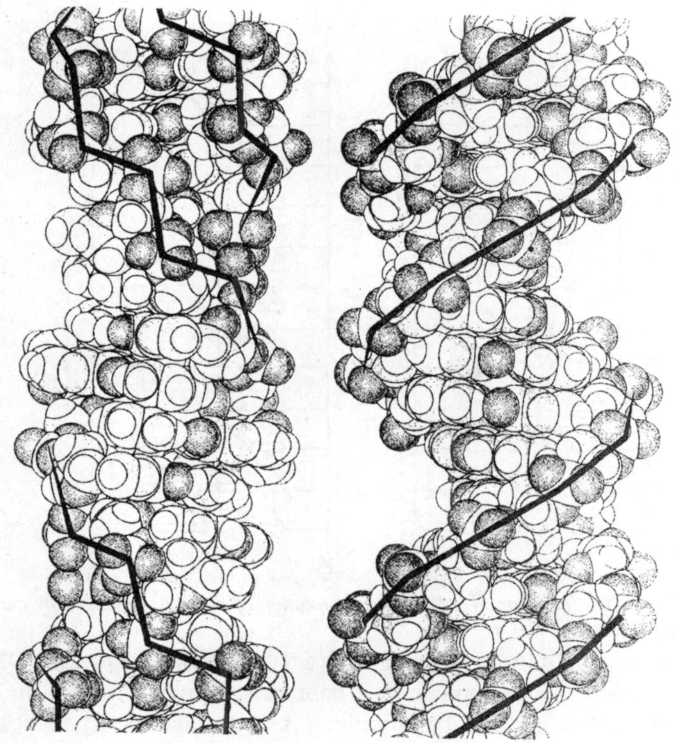

Fig. 1.12. Z (left) and B (right) DNA. The dark line connect phosphate groups.

anti configuration; the rotated configuration is called the *syn* configuration.)

Originally, it was thought that Z DNA would not prove of interest to biologists because it required very high salt concentrations to become stable. However, it was found that Z DNA can be stabilized in physiologically normal conditions if methyl groups are added to the cytosines. Z DNA may be involved in regulating gene expression in eukaryotes..

DNA Replication—Process

In their 1953 paper, Watson and Crick hinted that the replication of the double helix could take place as the DNA unwinds, so that each strand would form a new double helix by acting as a *template* for a newly synthesized strand. For example, when a double helix is unwound at an adenine-thymine (A-T) base pair, one unwound strand would carry A and the other would carry T. During replication, the A

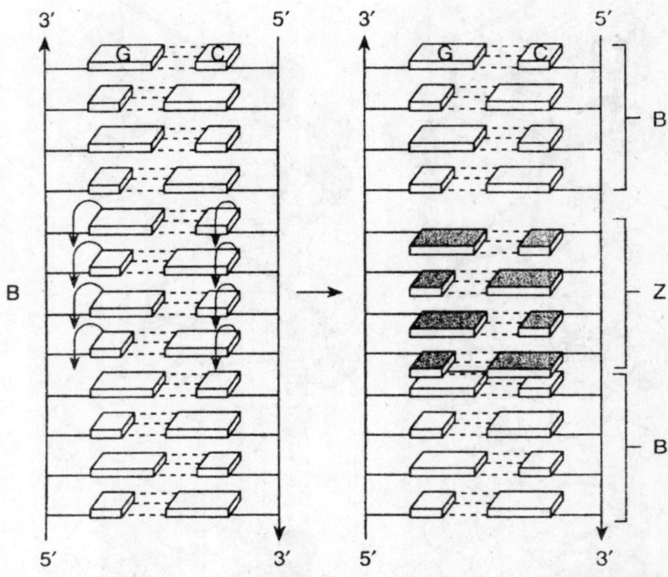

Fig. 1.13. B DNA converts to Z DNA by the rotation of bases as indicated by curved arrows.

in the template DNA would pair with a T in a newly replicated DNA strand, giving rise to another A-T base pair. Similarly, the T in the other template strand would pair with an A in the other newly replicated strand, giving rise to another A-T base pair. Thus, one A-T base pair in one double helix would result in two A-T base pairs in two double helices. This process would repeat at every base pair in the double helix of the DNA molecule.

This mechanism is called *semiconservative* replication because, although the entire double helix is not conserved in replication, each strand is. Every daughter DNA molecule has an intact template strand and a newly replicated strand. This is not the only way that replication could occur. The alternative methods are *conservative* and *dispersive*. In conservative replication, in which the whole original double helix acts as a template for a new one, one daughter molecule would consist of the original parental DNA, and the other daughter would be totally new DNA. In dispersive replication, some parts of the original double helix are conserved, and some parts are not. Daughter molecules would consist of part template and part newly synthesized DNA. In reality, the dispersive category is the all-inclusive "other" category, including any possibility other than conservative and semiconservative replication.

Meselson and Stahl Experiment

In 1958, M. Meselson and F. Stahl reported the results of an experiment designed to determine the mode of DNA replication. Some historians and philosophers of science consider this the most elegant scientific experiment ever designed. Meselson and Stahl grew *E. coli* in a medium containing a heavy isotope of nitrogen, ^{15}N. (The normal form of nitrogen is ^{14}N.) After growing for several generations on the ^{15}N medium, the DNA of *E. coli* was denser. The researchers determined the density of the strands using a technique known as *density-gradient centrifugation*. In this technique, a cesium chloride (CsCl) solution is spun in an ultracentrifuge at high speed for several hours. Eventually an equilibrium arises between centrifugal force and diffusion, so that a density gradient is established in the tube with an increasing concentration of CsCl from top to bottom. If DNA (or any other substance) is added, it concentrates and forms a band in the tube at the point where its density is the same as that of the CsCl. If several types of DNA with different densities are added, they form several bands. The bands are detectable under ultraviolet light at a wavelength of 260 nm (nanometers), which nucleic acids absorb strongly.

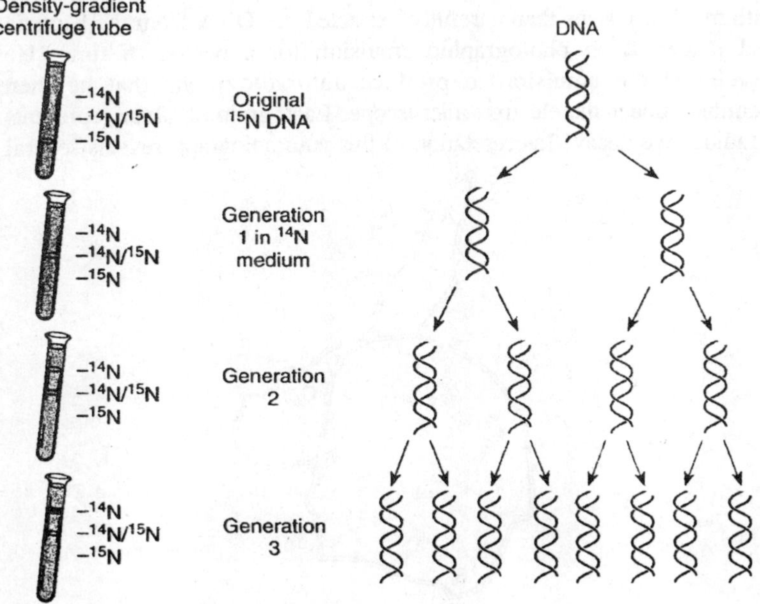

Fig. 1.14. The Meselson and Stahl experiment to determine the mode of replication of DNA.

Meselson and Stahl transferred the bacteria with heavy (^{15}N) DNA to a medium containing only ^{14}N. The new DNA, replicated in the ^{14}N medium, was intermediate in density between light (^{14}N) and heavy (^{15}N) DNA, because the replication was semiconservative. If replication had been conservative, two bands would have appeared at the first generation of replication—an original ^{15}N DNA and a new ^{14}N double helix. And, throughout the experiment, if the method of replication had been conservative, the original DNA would have continued to show up as a ^{15}N band. If the method of replication had been dispersive, various multiple-banded patterns would have appeared, depending on the degree of dispersiveness.

Autoradiographic Demonstration of DNA Replication

In 1963, J. Cairns used *autoradiography* to verify the semiconservative method of replication photographically. This technique makes use of the fact that radioactive atoms expose photographic film. The visible silver grains on the film can then be counted to provide an estimate of the quantity of radioactive material present. Cairns grew *E. coli* bacteria in a medium containing radioactive thymine, a component of one of the DNA nucleotides. The radioactivity was in tritium (^3H). Cairns then carefully extracted the DNA from the bacteria and placed it on photographic emulsion for a period of time. He developed the emulsion to produce autoradiographs that he then examined under the electron microscope. Each grain of silver represents a radioactive decay. Interpretation of this autoradiograph reveals several

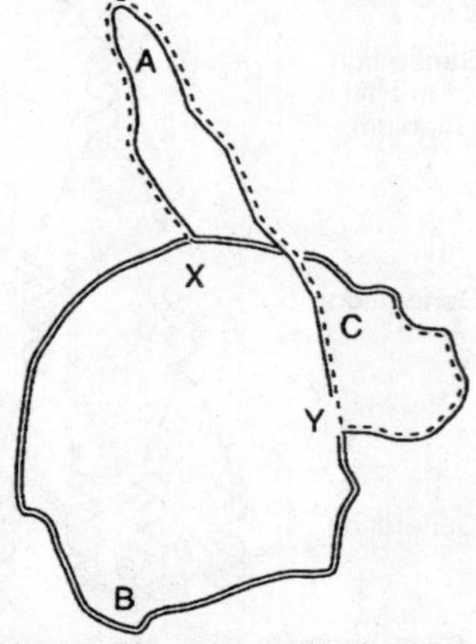

Fig. 1.15. Diagram has labels on the three segments, A, B, and C, created by the existence on two forks, X and Y, in the DNA.

points. The first, known at the time, is that the *E. coli* DNA is a circle. The second point is that the DNA is replicated while maintaining the integrity of the circle. That is, the circle does not appear to break during the process of DNA replication; an intermediate *theta structure* forms (topologically similar in shape to the Greek letter theta, θ). Third, replication of the DNA seems to be occurring at one or two moving *Y-junctions* in the circle, which further supports the semiconservative mode of replication. The DNA is unwound at a given point, and replication proceeds at a Y-junction, in a semiconservative manner, in one or both directions.

The steps by themselves do not support either a unidirectional or a bidirectional mode of replication. That is, a theta structure will develop if either one or both Y-junctions is active in replication. But with autoradiography, it is possible to determine whether new growth is occurring in only one or in both directions.

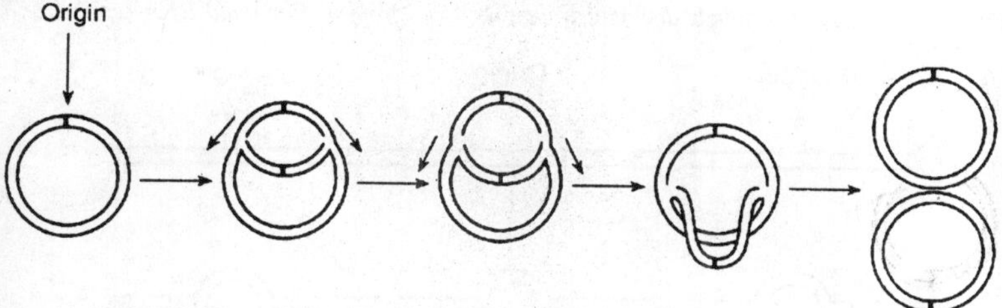

Fig. 1.16. Observable stages in the DNA replication of a circular chromosome, assuming bidirectional DNA synthesis. The intermediate figures are called theta structures.

In some cases, radioactivity was not applied to the cell until DNA replication had already begun. In these cases, the radioactive label appeared after the theta structure had already begun forming. By counting silver grains in autoradiographs, Cairns found growth to be bidirectional. Both autoradiographic and genetic analysis have subsequently verified this finding.

In eukaryotes, the DNA molecules (chromosomes) are larger than in prokaryotes and are not circular; there are also usually multiple sites for the initiation of replication. Thus, each eukaryotic chromosome is composed of many replicating units, or *replicons*—stretches of DNA with a single origin of replication. In comparison, the *E. coli* chromosome is composed of only one replicon. In eukaryotes, these replicating units form "bubbles" (or "eyes") in the DNA during replication.

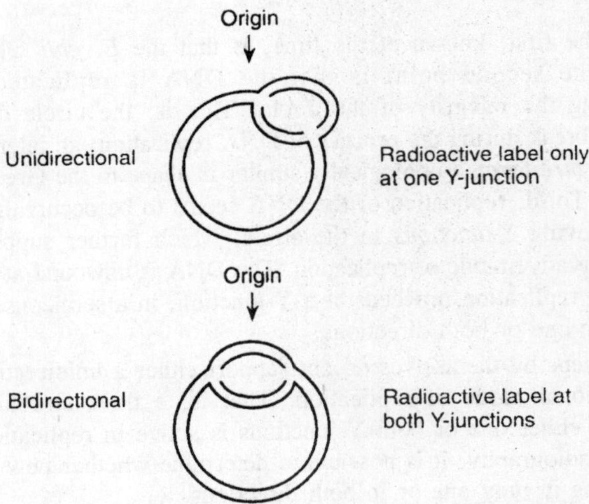

Fig. 1.17. Radioactive labels distinguish unidirectional from bidirectional DNA replication.

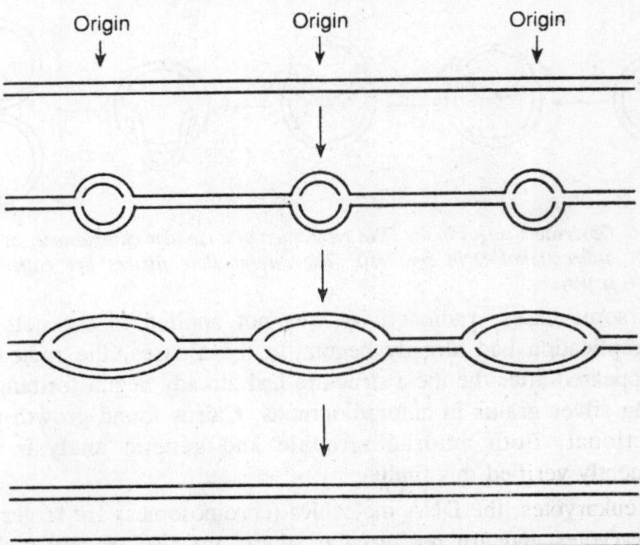

Fig. 1.18. Formation of bubbles (eyes) in eukaryotic DNA because of multiple DNA synthesis sites of origin.

DNA Replication—Enzymology

Let us turn now to the details of the processes that take place during DNA replication. Like virtually all metabolic processes, DNA replication is under the control of enzymes. The evidence for the details we describe comes from physical, chemical, and biochemical studies of enzymes and nucleic acids and from the analysis of mutations that influence the replication processes. More recent techniques of recombinant DNA technology and nucleotide sequencing have allowed us to determine the nucleotide sequences of many of these key regions in DNA and RNA. We will look first at *E. coli*.

There are three major enzymes that will polymerize nucleotides into a growing strand of DNA in *E. coli*. These enzymes are *DNA polymerase* I, II, and III. DNA polymerase I, discovered by Arthur Kornberg, who subsequently won the Nobel Prize for his work, is primarily utilized in filling in small DNA segments during replication and repair processes. DNA polymerase II can serve as an alternative repair polymerase; it can also replicate DNA if the template is damaged. DNA polymerase III is the primary polymerase during normal DNA replication.

In the simplest model of DNA replication, new nucleotides would be simultaneously added, according to the rules of complementarity, on both strands of newly synthesized DNA at the replication fork as the DNA opens up. But a problem exists, created by DNA's antiparallel nature; the two strands of a DNA double helix run in opposite directions. Going in one direction on the duplex, for example, one strand is a 5' → 3' strand, whereas the other is a 3' → 5' strand. These directions refer to the numbering of carbon atoms across the sugar. Since DNA replication involves the formation of two new antiparallel strands with the old single strands as templates, one new strand would have to be replicated in the 5' → 3' direction and the other in the 3' → 5' direction.

However, all the known polymerase enzymes add nucleotides in only the 5' → 3' direction. That is, the polymerase catalyzes a bond between the first 5'-PO_4 group of a new nucleotide and the 3'-OH carbon of the last nucleotide in the newly synthesized strand. The polymerases cannot create the same bond with the 5' phosphate of a nucleotide already in the DNA and the 3' end of a new nucleotide. Thus, the simple model needs some revision.

Continuous and Discontinuous DNA Replication

Autoradiographic evidence leads us to believe that replication occurs simultaneously on both strands. *Continuous replication* is, of course, possible on the 3' → 5' template strand, which begins with the necessary 3'-OH *primer*. (Primer is double-stranded DNA—or, as

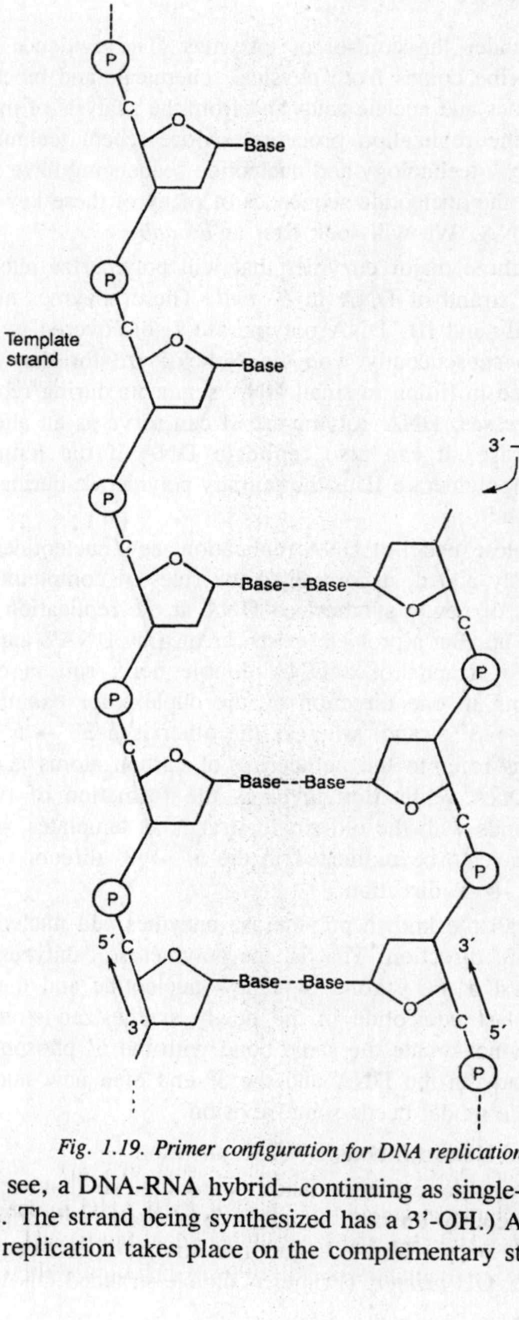

Fig. 1.19. *Primer configuration for DNA replication.*

we shall see, a DNA-RNA hybrid—continuing as single-stranded DNA template. The strand being synthesized has a 3'-OH.) A *discontinuous* form of replication takes place on the complementary strand, where it occurs in short segments, moving backward, away from the Y-junction. These short segments, called *Okazaki fragments* after R. Okazaki, who first saw them, average about 1,500 nucleotides in prokaryotes and 150 in eukaryotes. The strand synthesized continuously is referred to as the *leading strand*, and the strand synthesized discontinuously is referred to as the lagging strand.

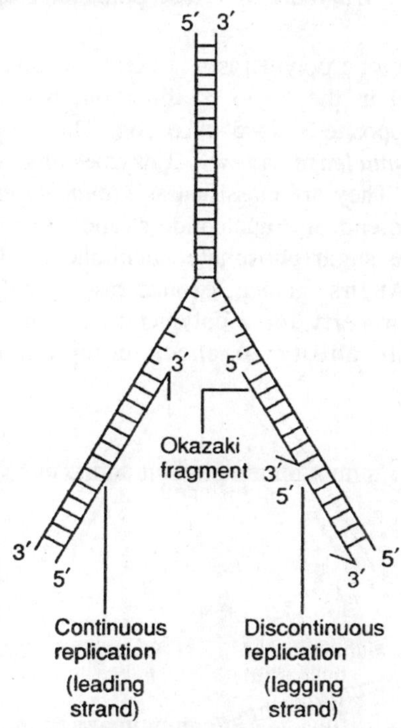

Fig. 1.20. Discontinuous model for DNA replication.

Once initiated, continuous DNA replication can proceed indefinitely. DNA polymerase III on the leading- strand template has what is called high *processivity*: once it attaches, it doesn't release until the entire strand is replicated. Discontinuous replication, however, requires the repetition of four steps: primer synthesis, elongation, primer removal with gap filling, and ligation.

Primer synthesis and elongation

To synthesize Okazaki fragments, a primer must be created *de novo* (Latin: from the beginning). None of the DNA polymerases can create that primer. Instead, *primase*, an RNA polymerase coded for by the *dnaG* gene, creates the primer, ten to twelve nucleotides, at the site of Okazaki fragment initiation. The result is a short RNA primer that provides the free 3'-OH group that DNA polymerase III needs in order to synthesize the Okazaki fragment. DNA polymerase III continues until it reaches the primer RNA of the previously synthesized Okazaki fragment. At that point, it stops and releases from the DNA.

All three prokaryotic polymerases not only can add new nucleotides to a growing strand in the 5' → 3' direction, but also can remove nucleotides in the opposite 3' → 5' direction. This property is referred to as 3' → 5' *exonuclease activity*. Enzymes that degrade nucleic acids are nucleases. They are classified as *exonucleases* if they remove nucleotides from the end of a nucleotide strand or as *endonucleases* if they can break the sugar-phosphate backbone in the middle of a nucleotide strand. At first glance, exonuclease activity seems like an extremely curious property for a polymerase to have—curious unless we think about its ability to check complementarity. If the

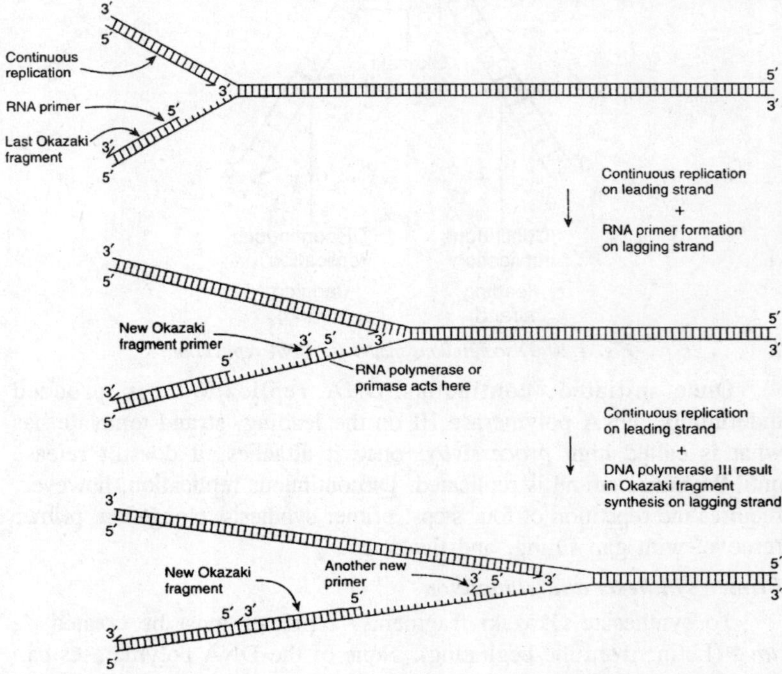

Fig. 1.21. Primer formation and elongation create an Okazaki fragment during discontinuous DNA replication.

complementarity is improper, meaning that the wrong nucleotide has been inserted, the polymerase can remove the incorrect nucleotide, put in the proper one, and continue on its way. This is known as the *proofreading* function of DNA polymerase. In addition, exonuclease activity can remove the RNA primers of Okazaki fragments.

Primer removal with gap filling

DNA polymerase I is a polymerase when it adds nucleotides, one at a time, and an exonuclease when it removes nucleotides one at a

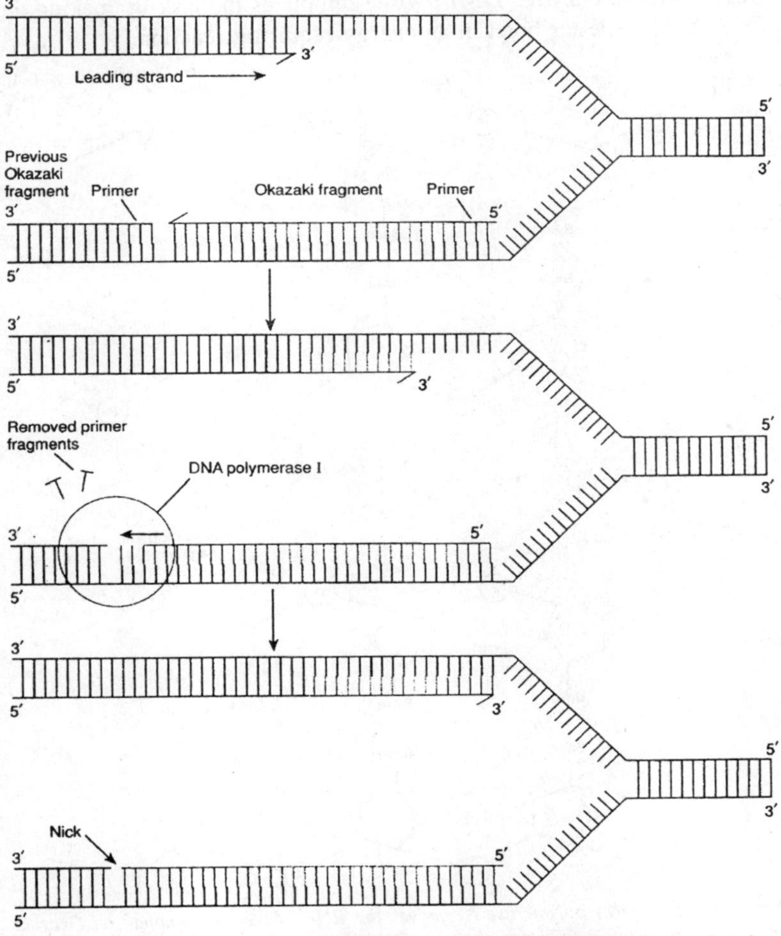

Fig. 1.22. The completion of an Okazaki fragment requires that DNA polymerase I replace the DNA primer base by base with DNA nucleotides.

time. To complete the Okazaki fragment, DNA polymerase I acts in both capacities. (DNA polymerase I mutants cannot properly connect Okazaki fragments.) DNA polymerase I completes the Okazaki fragment by removing the previous RNA primer and replacing it with DNA nucleotides. When DNA polymerase I has completed its nuclease and polymerase activity, the two previous Okazaki fragments are almost complete. All that remains is for a single phosphodiester bond to form.

Ligation

DNA polymerase I cannot make the final bond to join two Okazaki fragments. An enzyme, ***DNA ligase***, completes the task by making the final phosphodiester bond in an energy-requiring reaction.

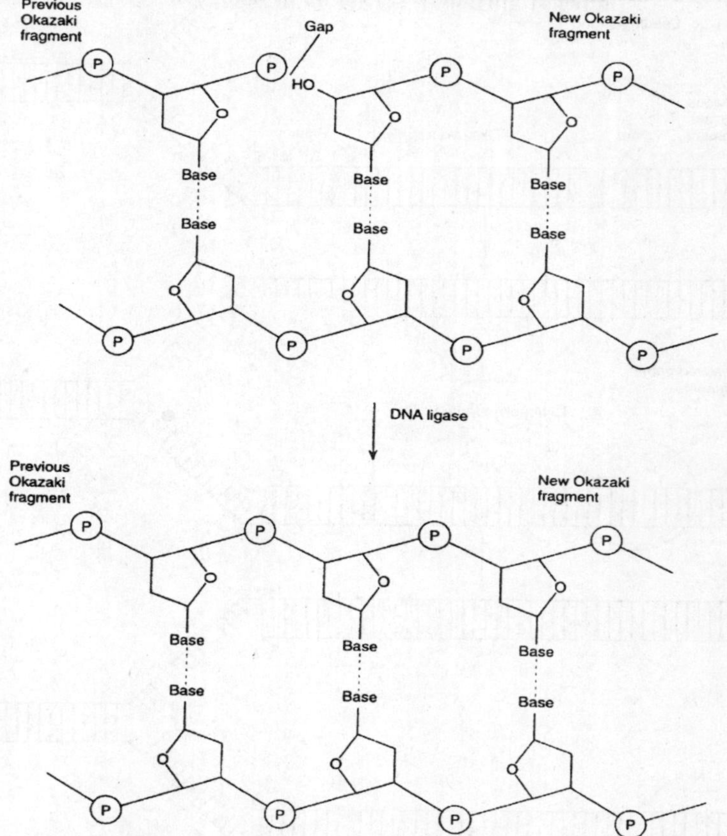

Fig. 1.23. After DNA polymerase I removes the RNA primer to complete an Okazaki fragment, a final gap remains. DNA ligase closes it.

A question of evolutionary interest is why RNA is used to prime DNA synthesis. Why not use DNA directly and avoid the exonuclease and resynthesis? Probably, making use of RNA primers lowers the error rate of DNA replication. That is, priming is an inherently error-prone process since nucleotides are initially added without a stable primer configuration. To prevent long-term errors in the DNA, an RNA primer is put in that can later be recognized and removed. Resynthesis by polymerase I is in a much more stable primer configuration (a long primer) and thus makes very few errors.

Another question of evolutionary interest is why DNA synthesis cannot take place in the 3' → 5' direction. Probably, the answer has to do with proofreading and the exonuclease removal of mismatched nucleotides. When an incorrect nucleotide is found and removed, the next nucleotide brought in, in the 5' → 3' direction, has a triphosphate end available to provide the energy for its own incorporation. Consider what would happen if the polymerase were capable of adding nucleotides in the opposite direction. The energy for the phosphodiester bond would be coming from the triphosphate already attached in the growing 3' → 5' strand. Then, if an error in complementarity were detected and the polymerase removed the most recently added nucleotide from the 3' → 5' strand, the last nucleotide in the double helix would no longer have a triphosphate available to provide energy for the diester bond with the next nucleotide. Continued polymerization would thus require additional enzymatic steps to provide the energy needed for the process to continue. This could stop or slow the process down considerably. As it is, the process incorporates about four hundred nucleotides per second with an error rate of about one incorrect pairing per 10^9 bases.

Origin of DNA Replication

Each replicon (e.g., the *E. coli* chromosome, or a segment of a eukaryotic chromosome with an origin of replication) must have a region where DNA replication initiates. In *E. coli*, this region is referred to as the genetic locus *oriC;* it occurs at map location 84 minutes. For DNA replication to begin, several steps must occur. First, the appropriate initiation proteins must recognize the specific origin site. Then the site must be opened and stabilized. And, finally, a replication fork must be initiated in both directions, involving continuous and discontinuous DNA replication. Although most of the proteins involved are known, there are still a few gaps in our knowledge.

OriC, the origin of replication in *E. coli*, is about 245 base pairs long and is recognized by *initiator proteins*. These proteins, the product

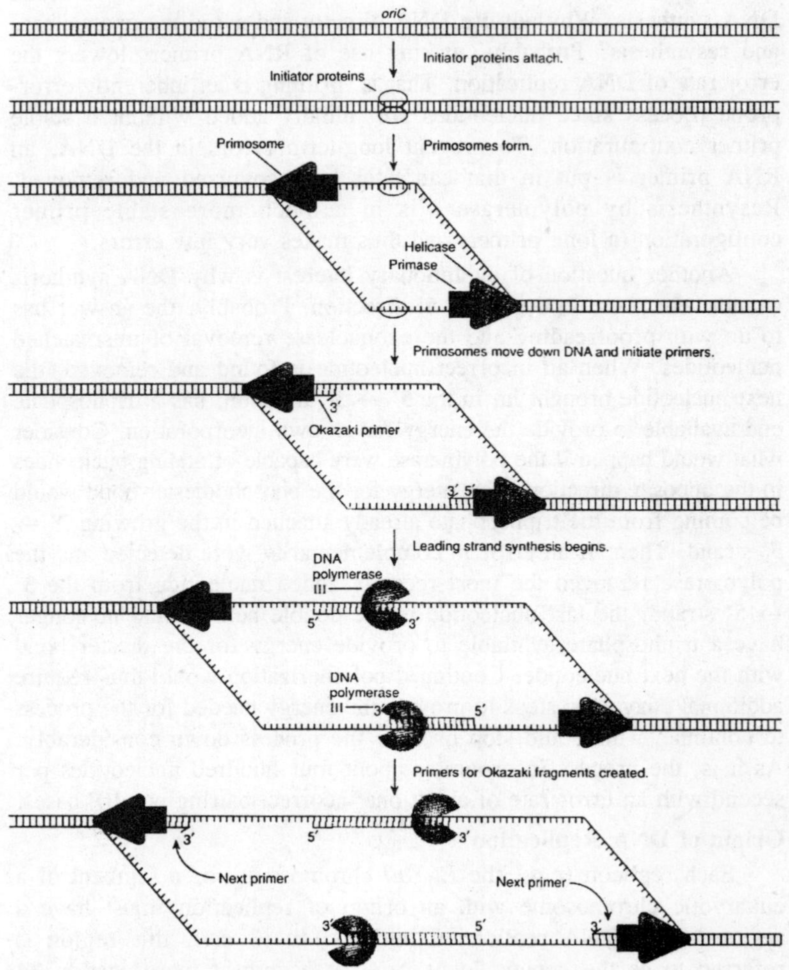

Fig. 1.24. Events at the origin of DNA replication in E. coli.

of the *dnaA* locus, open up the double helix. (Other DNA-binding proteins are also involved here.) The initiator proteins then take part in the attachment of DNA *helicase*, the product of the *dnaB* gene, which unwinds DNA at the Y-junction. Helicase is then responsible for recruiting (binding) the rest of the proteins that form the replication initiation complex. First is primase, which creates RNA primers. Together, the helicase and primase comprise a *primosome*, attached to the lagging-strand template. As the primosomes move along, they create

RNA primers that DNA polymerase III uses to initiate leading-strand synthesis. As primers are being laid down on the lagging-strand template, Okazaki fragment synthesis begins, and Y-junction activity then proceeds as outlined earlier.

DNA polymerase III *holoenzyme* is a very large protein composed of ten subunits. Three of the subunits, α, ε, and θ, form the polymerization core, with both 5' → 3' polymerase activity and 3' → 5' exonuclease activity. One subunit, the β subunit, is a "processivity clamp." As a dimer (two identical copies attached head to tail), the protein forms a "doughnut" around the DNA so it can move freely on the DNA. When it is attached to the core enzyme, the polymerase is held tightly to the DNA and shows high processivity: the leading strand is usually synthesized entirely without the enzyme leaving the template. The remaining subunits are involved in processivity control and replisome formation. They allow the polymerase to move off and on the DNA of the lagging-strand template as Okazaki fragments are completed (a process known as *polymerase cycling*).

Eukaryotes have evolved at least nine DNA polymerases, named DNA polymerase α, β, γ, δ, ε, ζ, η, θ, and τ. DNA polymerase seems to be the major replicating enzyme in eukaryotes, forming replisomes as in *E. coli*. In eukaryotes, the polymerase α-primase complex adds the Okazaki fragment primers, first adding an RNA primer and then a short length of DNA nucleotides. Polymerase ε may be involved in repair or in normal DNA replication, as is polymerase δ. DNA polymerase γ appears to replicate mitochondrial DNA. The remaining polymerases are probably involved in DNA repair, with polymerase β being the major repair polymerase, as polymerase I is in *E. coli*. Several of the polymerases most likely both replicate and repair DNA.

Eukaryotes also have a clamp-loader complex, called replication factor C, and a six-unit clamp called the proliferating cell nuclear antigen. The RNA primers are removed during Okazaki fragment completion (maturation) by mechanisms similar to those in prokaryotes. In eukaryotes, RNAase enzymes remove the RNA primers in Okazaki fragments; a repair polymerase fills gaps; and a DNA ligase forms the final seal. Helicases, topoisomerases, and single-strand binding proteins play roles similar to those they play in prokaryotes. The completion of the replication of linear eukaryotic chromosomes involves the formation of specialized structures at the tips of the chromosomes. Thus, all of the enzymatic processes are generally the same in prokaryotes and eukaryotes. DNA replication developed in prokaryotes and was refined as prokaryotes evolved into eukaryotes.

T. Steitz and his colleagues have done much X-ray crystallography work that has given us an excellent look at the structure of a polymerase. (Most work has actually been done on a fragment of DNA polymerase I called the *Klenow fragment*.) The enzyme is shaped like a cupped right hand with enzymatic activity taking place in two places, separated by a distance of about two to three nucleotides. It is proposed that when the polymerization site senses a mismatch, the DNA is moved so that the 3' end enters the exonuclease site, where the incorrect nucleotide residue is then cleaved. Polymerization then continues. There may be a general mode of polymerase action among diverse polymerases.

The replication of the *E. coli* chromosome may be controlled by the methylation state of several sequences within *oriC*. Certain enzymes add methyl groups to specific DNA bases, and the presence or absence of these methyl groups can serve as signals to other enzymes.

Events at the Y-Junction

We now have the image of DNA replication proceeding as a primosome moves along the lagging-strand template, opening up the DNA (helicase activity), and creating RNA primers (primase activity) for Okazaki fragments. One DNA polymerase III moves along the leading-strand template, generating the leading strand by continuous DNA replication, whereas a second DNA polymerase III moves backward, away from the Y-junction, creating Okazaki fragments. *Single-strand binding proteins* (ssb proteins) keep single-stranded DNA stabilized (open) during this process, and DNA polymerase I and ligase connect Okazaki fragments.

This simple picture is slightly complicated by the fact that the lagging- and leading-strand synthesis is coordinated. B. Alberts suggested an explanation: the *replisome* model, in which both copies of DNA polymerase III are attached to each other and work in concert with the primosome at the Y-junction. According to this model, a single replisome, consisting of two copies of DNA polymerase III, a helicase, and a primase, moves along the DNA. The leading-strand template is immediately fed to a polymerase, whereas the lagging-strand template is not acted on by the polymerase until an RNA primer has been placed on the strand, meaning that a long (fifteen-hundred base) single strand has been opened up. As the replisome moves along, another single-stranded length of the lagging-strand template forms. At about

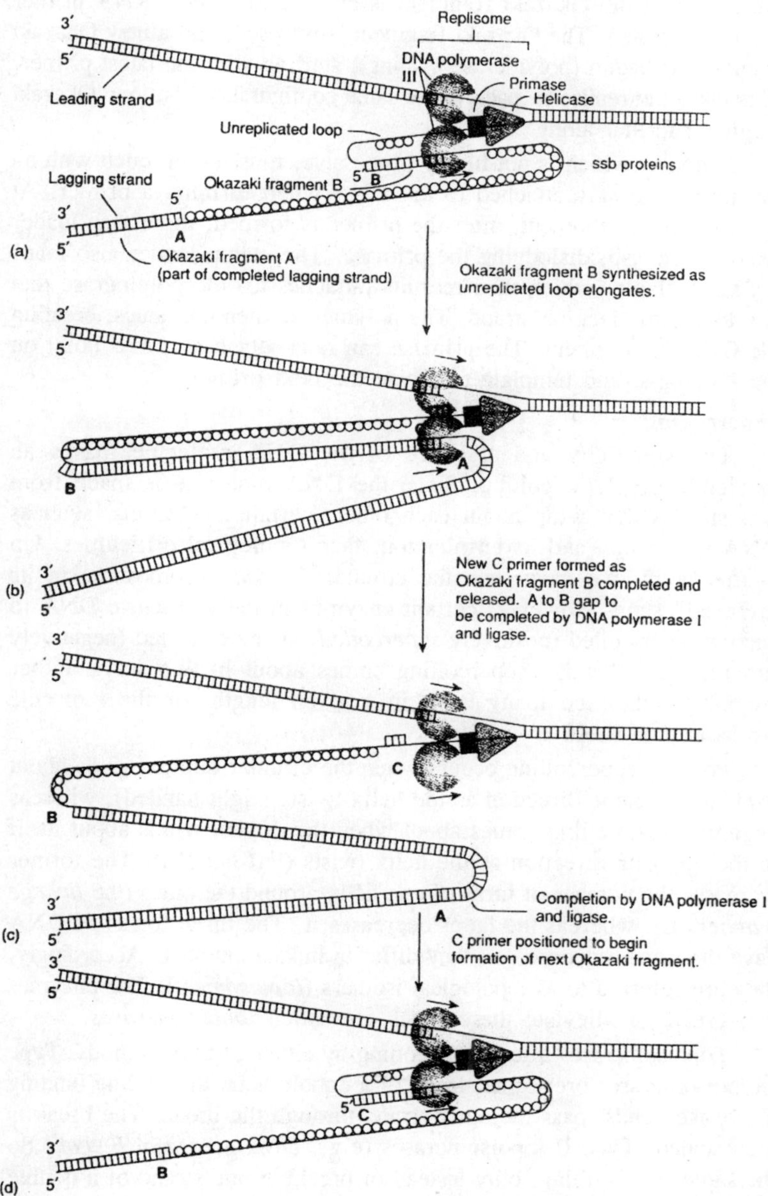

Fig. 1.25. The replisome, which consists of two DNA polymerase III holoenzyme and a primosome coordinates replication at the Y-junction.

the time that the Okazaki fragment is completed, a new RNA primer has been created. The Okazaki fragment is released, and a new Okazaki fragment is begun (polymerase cycling), starting with the latest primer. This takes the replisome back to the same configuration, but one Okazaki fragment farther along.

Primase, which is not highly processive, must be in touch with an ssb protein to stay attached to the DNA when forming a primer. At the appropriate moment, after the primer is formed, the clamp loader contacts the ssb, dislodging the primase. The clamp loader also loads a sliding clamp, which then recruits (attaches to) the polymerase that is creating the lagging strand. The polymerase then continues, creating the Okazaki fragment. The primase can later attach at a new point on the lagging-strand template to create the next primer.

Supercoiling

The simplicity and elegance of the DNA molecule masks an inevitable problem: coiling. Since the DNA molecule is made from two strands that wrap about each other, certain operations, such as DNA replication and its termination, face topological difficulties. Up to this point, we have seen the circular *E. coli* chromosome in its "relaxed" state. However, certain enzymes in the cell cause DNA to become over-coiled (positively *supercoiled*) or under-coiled (negatively supercoiled). Positive supercoiling comes about in two ways: either the DNA takes too many turns in a given length, or the molecule wraps around itself.

Positive supercoiling occurs when the circular duplex winds about itself in the same direction as the helix twists (right-handed), whereas negative supercoiling comes about when the duplex winds about itself in the opposite direction as the helix twists (left-handed). The former increases the number of turns of one helix around the other (the *linkage number*, L), whereas the latter decreases it. The three forms of DNA have the same sequence, yet they differ in linkage number. Accordingly, they are referred to as topological isomers (*topoisomers*). The enzymes that create or alleviate these states are called *topoisomerases*.

Topoisomerases affect supercoiling by either of two methods. Type I topoisomerases break one strand of a double helix and, while binding the broken ends, pass the other strand through the break. The break is then sealed. Type II topoisomerases (e.g., *DNA gyrase* in *E. coli*) do the same sort of thing, only instead of breaking one strand of a double helix, they break both and pass another double helix through the temporary gap. Four topoisomerases are active in *E. coli*, with somewhat

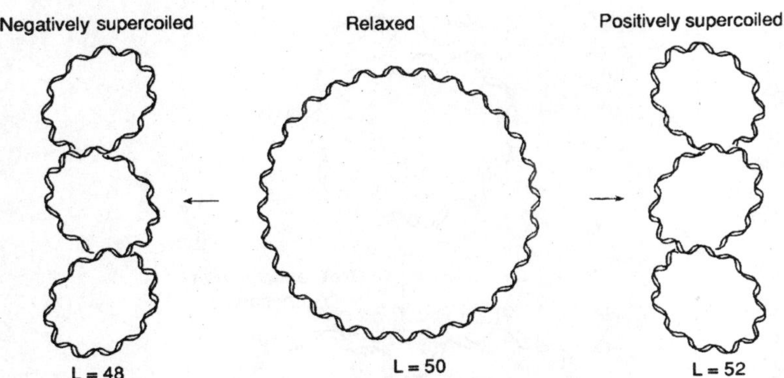

Fig. 1.26. Positive and negative supercoils.

confusing nomenclature: topoisomerases I and III are type I; topoisomerases II and IV are type II.

As DNA replication proceeds, positive supercoiling builds up ahead of the Y-junction. This is eliminated by topoisomerases that either create negative supercoiling ahead of the Y-junction in preparation for replication or alleviate positive supercoiling after it has been created.

Termination of Replication

The termination of the replication of a circular chromosome presents no major topological problems. At the end of the theta-structure replication, both Y-junctions have proceeded around the molecule. The region of termination on the *E. coli* chromosome, the terminus region, is 180 degrees from *oriC* on the circular chromosome, between minutes 28 and 36. There are six terminator sites (*Ter*); three arrest the Y-junction from the left, and three arrest the one from the right when bound by a termination protein, the protein product of the *tus* gene. (*Tus* stands for terminus utilization substance; each *Ter* site is about twenty base pairs.) One interesting aspect of the termination of *E. coli* DNA replication is that the cells are viable even if the whole terminator region is deleted. There are fewer viable cells and some growth problems, but in general, *E. coli* can successfully terminate DNA replication even without formal termination sites. A topoisomerase, topoisomerase IV, then releases the two circles, and DNA polymerase I and ligase close them up.

DNA Partitioning in E. coli

As we have discussed the processes that partition eukaryotic chromosomes between daughter cells during mitosis and meiosis. Until very recently, geneticists believed that the partitioning of the *E. coli*

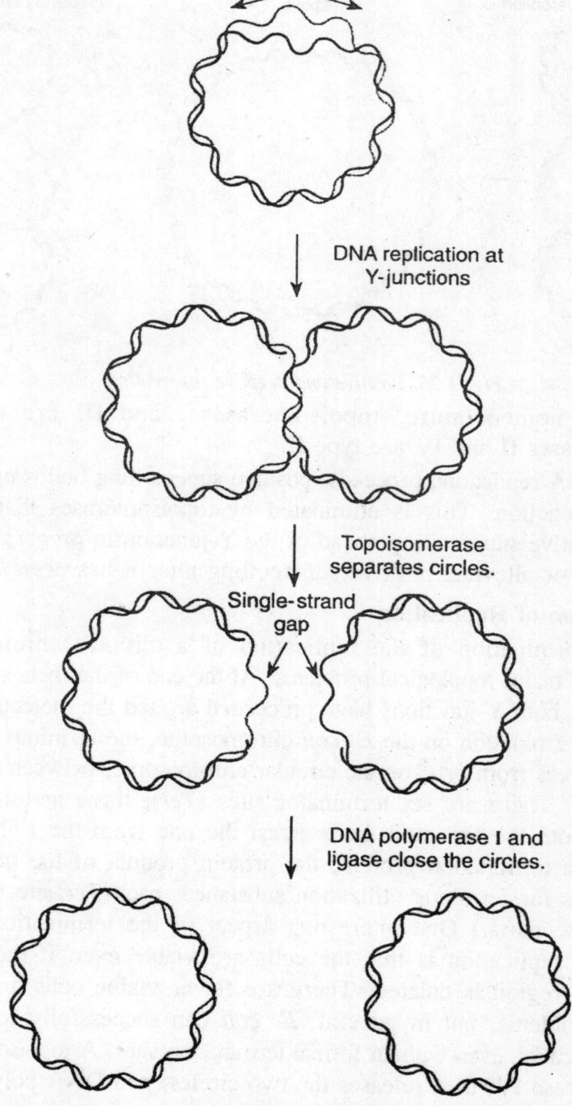

Fig. 1.27. The replication of circular DNA terminates when topoisomerase separates the circles and DNA polymerase I and ligase close the gaps in each circle.

chromosome was a passive process, unlike that in eukaryotes. Now, however, we know that more complexity is involved in *E. coli* DNA partitioning. When DNA replication begins, the newly replicated origins of replication are segregated to opposite ends of the bacterial cell, acting as centromeres do. A ring of proteins, the products of the *FtsZ* gene, form a ring at the middle of the cell and begin to create the septum that will divide the cell into two. The full complexity involved in *E. coli* chromosomal partitioning should be uncovered in the near future.

Replication Structures

The *E. coli* model of DNA replication that we have presented here is by way of the intermediate theta-structure. Two other modes of replication occur in circular chromosomes: rolling-circle and D-loop.

Rolling-Circle Model

In the *rolling-circle* mode of replication, a nick (a break in one of the phosphodiester bonds) is made in one of the strands of the circular DNA, resulting in replication of a circle and a tail. This form of replication occurs in the F plasmid or *E. coli* Hfr chromosome during conjugation. The F^+ or Hfr cell retains the circular daughter while passing the linear tail into the F^- cell, where replication of the tail takes place. Several phages also use this method, filling their heads (protein coats) with linear DNA replicated from a circular parent molecule.

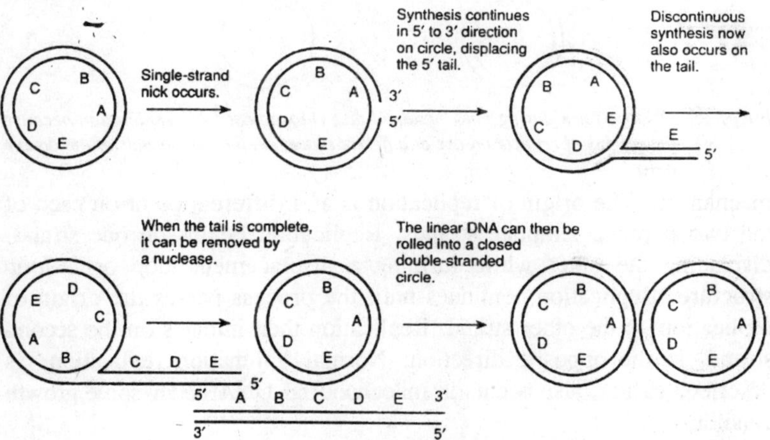

Fig. 1.28. Rolling-circle model of DNA replication.

D-Loop Model

Chloroplasts and mitochondria (in eukaryotic cells) have their own circular DNA molecules that appear to replicate by a slightly different

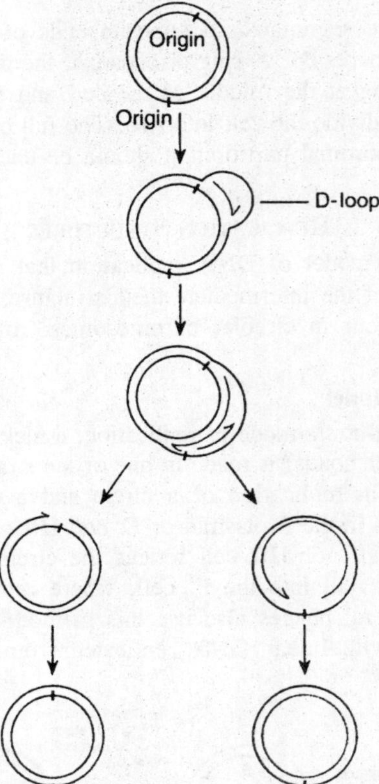

Fig. 1.29. D-loops form during mitochondrial and chloroplast DNA replication because the origins of replication are at different places on the two strands of the double helix.

mechanism. The origin of replication is at a different point on each of the two parental template strands. Replication begins on one strand, displacing the other while forming a displacement loop or *D-loop* structure. Replication continues until the process passes the origin of replication on the other strand. Replication then initiates on the second strand, in the opposite direction. Normal Y-junction replication, as described earlier, also occurs in mitochondrial DNA under some growth conditions.

Eukaryotic DNA Replication

As we saw earlier, linear eukaryotic chromosomes usually have multiple origins of replication, resulting in figures referred to as "bubbles" or "eyes". Multiple origins allow eukaryotes to replicate their larger quantities of DNA in a relatively short time, even though eukaryotic DNA replication is considerably slowed by the presence of histone proteins associated with the DNA to form chromatin. For example, the *E. coli* replication fork moves through about twenty-five thousand base pairs per minute, whereas the eukaryotic Y-junction moves through only about two thousand base pairs per minute. The number of replicons in eukaryotes varies from about five hundred in yeast to as many as sixty thousand in a diploid mammalian cell.

In budding yeast, a lower eukaryote that is often used as a model organism, DNA replication initiates at sites called *autonomously replicating sequences* (ARS). Each consists of a specific 11-base-pair sequence plus two or three additional short DNA sequences encompassing 100–200 base pairs. Six proteins form a complex that binds to this sequence, referred to as the *origin recognition complex* (ORC). These proteins seem to be bound all the time, and thus additional proteins are needed to initiate DNA replication. Some of these additional proteins are cyclin-dependent kinases, proteins involved in the control of the cell cycle. This makes sense because in eukaryotes, DNA replication can take place only once during the cell cycle, during the S phase. Thus, the initiation of DNA replication must be tightly controlled to avoid multiple replication of some or all replicons.

2

Biological Replication of Gene

Genetic material is transmitted from parent to offspring and from cell to cell. For this to occur, the genetic material must be copied. During this process, known as *DNA replication*, the original DNA strands are used as templates for the synthesis of new DNA strands This chapter will begin with a consideration of the structural features of the double helix that pertain to the replication process. Then we will examine how chromosomes are replicated within living cells: where does DNA replication begin, how does it proceed, and where does it end? At the molecular level, it is rather remarkable that the replication of chromosomal DNA occurs very quickly very accurately, and at the appropriate time in the life of the cell. For this to hap pen, many cellular proteins play vital roles. In this chapter, we will examine the mechanism of DNA replication and consider the functions of several proteins involved in the process.

Structural Overview of DNA Replication

We begin by recalling a few important structural features of the double helix, since they bear directly on the replication process. The double helix is composed of two DNA strands, and the individual building blocks of each strand are nucleotides. The nucleotides contain one of four bases: adenine, thymine, guanine, or cytosine. The double-stranded structure is held together by hydrogen bonding between the bases in opposite strands. A critical feature of the double helix structure is that it will only fit together if adenine hydrogen bonds with thymine and guanine hydrogen bonds with cytosine. This rule, known as the A—T/G—C rule or Chargaff's rule, is the basis for the complementarity of the base sequences in double-stranded DNA.

Another feature worth noting is that the strands within a double helix have an antiparallel alignment. This directionality is determined by the orientation of sugar residues within the sugar—phosphate backbone. If one strand is running in the 5' to 3' direction, the complementary strand is running in the 3' to 5' direction. The issue of directionality will be important later in this chapter, when we consider the function of the enzymes that synthesize new DNA strands. In this section, we will consider how the structure of the DNA double helix allows it to be replicated.

The Complementarity of Base Sequences Provides the Basis for DNA Replication

DNA replication relies on the complementarity of DNA strands according to the A—T/G—C rule. During the replication process, the two complementary strands of DNA come apart and serve as *template strands* for the synthesis of two new strands of DNA. After the template strands have separated, individual nucleotides can hydrogen bond with the template strands. This hydrogen bonding must obey the A—T/G—C rule. To complete the replication process, a covalent bond is formed between the phosphate on one nucleotide and the sugar on the previous nucleotide. The two newly made strands are referred to as the *daughter strands*, while the original strands are called the *parental strands*. Note that the base sequences are identical in both double-stranded molecules after replication. Therefore, DNA can be replicated so that both copies retain the same information as the original molecule.

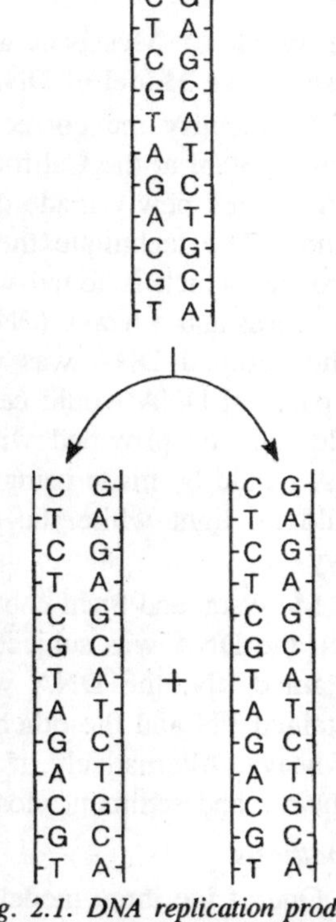

Fig. 2.1. DNA replication produces two copies of DNA with the same sequence as the original DNA molecule.

In the 1950s, Three Different Models were Proposed that Described the Net Result of DNA Replication

Scientists in the late 1950s had considered three different mechanisms that de scribed what the net result of DNA replication would be. The first is referred to as a *conservative* model. According to this hypothesis, both strands of parental DNA remain together following DNA replication. Therefore, the original arrangement of parental strands is completely conserved, and the two daughter strands are also together following replication. The second is called a *semiconservative* model. The double-stranded DNA is "half" conserved following the replication process. In other words, the double-stranded DNA contains one parental strand and one daughter strand. The third, *dispersive*, model proposes that segments of parental DNA and newly made DNA are interspersed in both strands following the replication process.

The Work of Meselson and Stahl in 1958 Supported the Semi-conservative Model of DNA Replication

To identify the correct model of those Matthew Meselson and Franklin Stahl at the California Institute of Technology experimentally distinguished newly made daughter strands from the original parental strands. The technique they used is called *heavy isotope labeling*. Nitrogen, which is found within the bases of DNA, occurs in a light (^{14}N) form and a heavy (^{15}N) form. Meselson and Stahl reasoned that if the parental DNA was made within cells that only have ^{15}N, all the parental DNA would be heavy Prior to DNA replication, the cells could then be provided with only ^{14}N so that all their newly made DNA would be made using ^{14}N Therefore, the daughter DNA strands would be light while the original parental DNA strands would be heavy.

Meselson and Stahl could then analyze the labeling of DNA strands when the DNA was subjected to centrifugation. If both DNA strands contained ^{14}N, the DNA would have a light density. If one strand contained ^{14}N and the other strand contained ^{15}N, the DNA would be half-heavy. Alternatively, if both strands contained ^{15}N, the DNA would be heavy and sediment closer to the bottom of the centrifuge tube.

Hypothesis

One of the three models may accurately describe the net result of DNA replication. The purpose of this experiment is to determine which of the three models is correct.

Testing the hypothesis

Starting material: A strain of *E. coli* that has been grown for many generations in the presence of ^{15}N. All of the nitrogen in the DNA is labeled with ^{15}N.

1. Add an excess of ^{14}N containing compounds to the bacterial cells so that all the newly made DNA will contain ^{14}N.
2. Incubate the cells for various lengths of time.
3. Lyse the cells by the addition of lysozyme and detergent, that disrupt the bacterial cell wall and cell membrane respectively.
4. Load the lysate onto a CsCl gradient that contains ethidium bromide. (Note: The a density of DNA is around 1.7 gm/cm^3, which is well isolated from other cellular macromolecules.)
5. Centrifuge the gradients until the DNA molecules reach their equilibrium densities.
6. DNA within the gradient can be observed with a UV light since the DNA is now stained with ethidium bromide, a UV sensitive dye.

Interpreting the data

After one complete generation (i.e, one round of DNA replication), all the DNA sedimented at a density that was half-heavy. These results are consistent with both the semiconservative and dispersive models. The conservative model had predicted two separate DNA types a light type and a heavy type. Since all the DNA had sedimented as a single band, this model was now disproved. According to the semi- conservative model, the replicated DNA would contain one original strand (a heavy strand) and a newly made daughter strand (a light strand). Likewise, in a dispersive model, there should have been all half-heavy DNA after one generation as well. To determine which of these two remaining models was correct, therefore, Meselson and Stahl had to investigate future generations.

After approximately two generations (precisely, 1.9 generations), there was a mixture of light DNA and half-heavy DNA. This observation was also consistent with the semiconservative mode of DNA replication, because some DNA molecules should contain all light DNA while other molecules should behalf-heavy. In a dispersive model, however, the DNA after two generations would have been 1/4-heavy. This is because a dispersive model predicts that the heavy nitrogen would be evenly dispersed among four strands, each strand thus containing 1/4 heavy nitrogen and 3/4 light nitrogen. However, this result was not obtained. Instead, the results of the Meselson and Stahl experiment provided compelling evidence in favor of the semiconservative model for DNA replication.

Prokaryotic DNA Replication

Thus far in this chapter, we have considered how complementarity underlies the ability of DNA to be copied. In addition, the experiments of Meselson and Stahl showed that DNA replication results in two double helices, each one containing an original parental strand and a newly made daughter strand. We will now turn our attention to how DNA replication actually occurs within living cells. Much re search has focused on DNA replication in the bacterium *E. coli*. The results of these studies have provided the foundation for our current understanding of DNA replication. The replication of the bacterial chromosome is a stepwise process in which many cellular proteins participate. In this section, we will follow this process from beginning to end.

Bacterial Chromosomes Contain a Single Origin of Replication

The site on the bacterial chromosome where DNA synthesis begins is known as the origin of replication. Bacterial chromosomes have a single origin of replication. The synthesis of new daughter strands is initiated within the origin and proceeds *bidirectionally* (in both directions) around the bacterial chromosome. This means that two *replication forks* move in opposite directions outward from the origin. Eventually, these replication forks meet each other on the opposite side of the bacterial chromosome to complete the replication process.

Replication is Initiated by the Binding of DnaA Protein to the Origin of Replication

Considerable research has focused on the origin of replication in *E. coli*. This origin has been named *oriC*. There are three types of DNA sequences within oriC that are functionally important: an A/T-rich region, dnaA box sequences, and GATC methylation sites. The third functional sequence, GATC methylation sites, will be discussed later when we consider the regulation of replication.

DNA replication is initiated by the binding of *DnaA proteins* to sequences within the origin known as *dnaA box sequences*. The dnaA box serves as a recognition site for the binding of the DnaA protein. Several DnaA proteins bind to the four dnaA boxes in *oriC* to initiate DNA replication. The DnaA protein has several important functions in DNA replication. First, it binds specifically to the dnaA boxes in

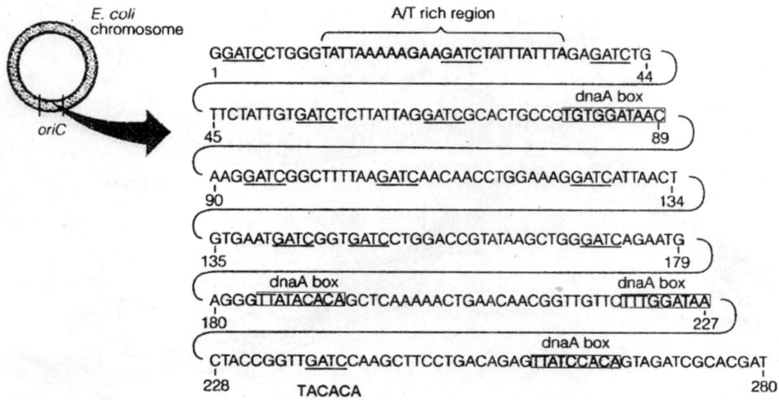

Fig. 2.2. The sequence of oriC in E.coli.

oriC, thereby recognizing this region as the single origin of replication. Next, DnaA proteins cause the A/T-rich region to denature into separate single- stranded regions. Since A/T base pairs form only two hydrogen bonds while GIG base pairs form three, it is easier to separate two DNA strands in an A/T-rich region.

Following denaturation of the A/T-rich region, the DnaA proteins, with the help of the DnaC protein, recruit *DNA helicase* (also known as Dna B helicase) to bind to this site. DNA helicases begin strand separation within the *oriC* region and continue to separate the DNA strands beyond the origin. These enzymes use the energy from ATP hydrolysis to catalyze the separation of the double-stranded parental DNA. DNA helicases bind to single-stranded DNA and travel along the DNA in a 5' to 3' direction to keep the replication fork moving. When a helicase encounters a double-stranded region, it breaks the hydrogen bonds between the two strands, thereby generating two single strands. The action of DNA helicases promotes the movement of the two replication forks outward from *oriC* in opposite directions.

DNA Strand Separation and the Synthesis of RNA Primers are Necessary before Daughter Strands of DNA can be Made

The action of DNA helicase is believed to generate positive supercoiling ahead of the replication fork. An enzyme known as *topoisomerase* (i.e., DNA gyrase) travels ahead of the helicase enzyme and alleviates this positive supercoiling.

After the two parental strands have been separated and the supercoiling relaxed, they must be kept that way until the complementary daughter strands have been made. The function of the

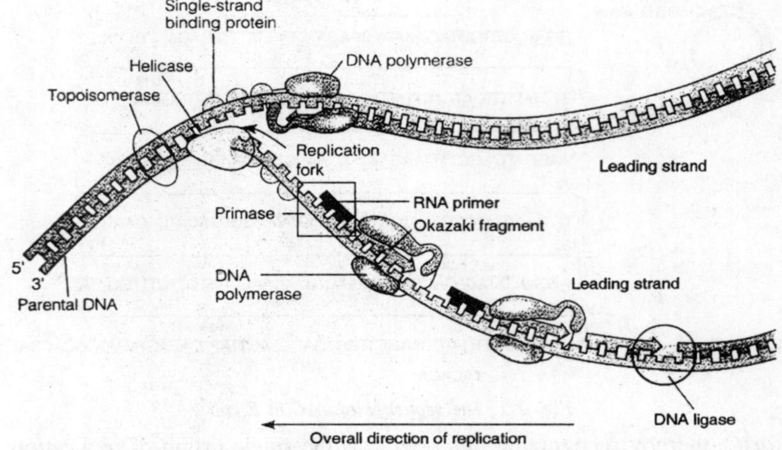

Fig. 2.3. Enzymology of DNA replication.

single-strand binding protein is to bind to both of the single strands of parental DNA and prevent them from re-forming a double helix. In this way, the bases within the parental strands are kept in an exposed condition that enables them to hydrogen bond with single nucleotides.

The next event in DNA replication involves the synthesis of a short strand of RNA (rather than DNA) called an *RNA primer*. This strand of RNA is synthesized by the linkage of ribonucleotides via an enzyme known as *primase*. These RNA strands start, or prime, the process of DNA replication. At a later stage in DNA replication, the RNA primers are removed.

DNA Polymerases Link Nucleotides to Synthesize the Daughter Strands

The enzymes known as *DNA polymerases* are responsible for covalently attaching nucleotides together to make new daughter strands. In *E. coli*, there are three distinct proteins that function as DNA polymerases, designated *polI*, *polII*, and *polIII*. *PolI* and *polIII* are involved in DNA replication, and *polII* is important in DNA repair. In DNA replication, there are two important features of DNA polymerases that may seem unusual:

1. DNA polymerases cannot link together the first two nucleotides in a new strand. Rather, they can only elongate a strand starting with a primer.
2. DNA polymerases can attach nucleotides only in the 5' to 3' direction, not in the 3' to 5' direction.

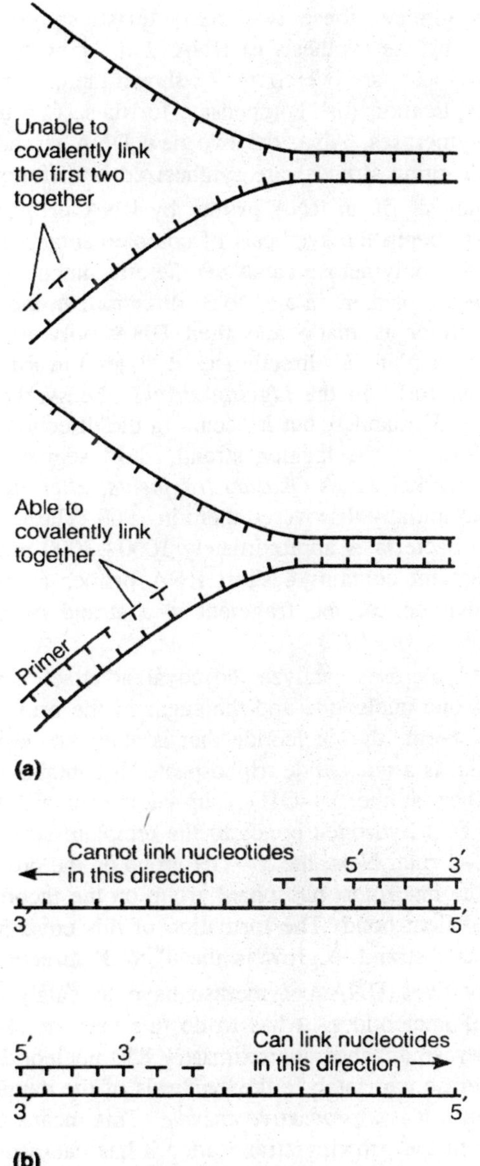

Fig. 2.4. Unusual features of DNA polymerase function. (a) DNA polymerase can only elongate a strand starting with a short RNA primer. (b) DNA polymerase can only attach nucleotide in a 5′ to 3′ direction.

At first glance, these two characteristics may seem to pose problems during the synthesis of DNA. Let's now turn back and see how these problems are overcome. As shown there, additional enzymes within the replication fork compensate for these two unusual features of DNA polymerases. Also, the two new DNA strands (namely, the leading and lagging strands) are synthesized in different ways.

The synthesis of an RNA primer by RNA primase allows DNA polymerase to begin the synthesis of complementary daughter strands of DNA. DNA polymerase catalyzes the attachment of nucleotides to the 3' end of the primer, in a 5' to 3' direction. In the *leading strand*, one RNA primer is made and then DNA polymerase can attach nucleotides in a 5' to 3' direction as it slides toward the opening of the replication fork. In the *lagging strand*, the synthesis of DNA is also in a 5' to 3' manner, but it occurs in the direction away from the replication fork. In the lagging strand, short segments of DNA are made; these are known as *Okazaki fragments*, after Reiji and Tuneko Okazaki, who initially discovered them in 1968. The length of Okazaki fragments in bacteria is approximately 1000—2000 nucleotides. Each Okazaki fragment contains a short RNA primer at the 5' end, and then the remainder of the fragment is a strand of DNA made by DNA polymerase (*polIII*).

DNA polymerases catalyze the covalent attachment between the phosphate in one nucleotide and the sugar in the previous nucleotide. Prior to this event, the nucleotide that is about to be attached to the growing strand is a nucleoside triphosphate. It contains three phosphate groups attached at the 5'—OH group of the sugar. The nucleoside triphosphate first hydrogen bonds to the template strand according to the A—T/G—C rule. Next, the 3'—OH group on the previous nucleotide reacts with the innermost phosphate group on the incoming nucleotide to form a covalent bond. The formation of this covalent bond causes the newly made strand to grow in the 5' to 3' direction.

Not only does DNA polymerase have to catalyze the covalent attachment of nucleotides, it has to do this very quickly. In *E. coli*, DNA polymerase attaches approximately 850 nucleotides per second! DNA polymerase can catalyze the synthesis of the daughter strands so quickly because it is a *processive enzyme*. This means that it does not dissociate from the growing strand after it has catalyzed the covalent joining of two nucleotides. Rather, it remains clamped to the DNA template strand and slides along the template as it catalyzes the synthesis of the daughter strand. Certain proteins, known as processivity factors, are responsible for this clamping phenomenon. For example, in *E. coli*, processivity factors called β (beta), τ (tau), and γ (gamma), which are subunits of *polIII*, are important in clamping the enzyme to the DNA template strand.

DNA Polymerase (polI) Removes the RNA Primers in the Lagging Strand and Fills in the Gap with DNA, and then DNA Ligase Links the Okazaki Fragments

To complete the synthesis of Okazaki fragments within the lagging strand, three additional events must occur: removal of the RNA primers, synthesis of DNA in the area where the primers have been removed, and the covalent attachment of adjacent fragments of DNA. In *E. coli*, the RNA primers are removed by the action of DNA polymerase (*polI*). This enzyme has a 5' to 3' exonuclease activity. This means that *polI* digests away the RNA primers in a 5' to 3' direction, leaving a vacant area. *PolI* can then fill in this region by synthesizing DNA there. After the gap has been completely filled in, there is still a bond missing between the last nucleotide added by *polI* and the first nucleotide in the next DNA fragment. An enzyme known as *DNA ligase* catalyzes a covalent bond between these adjacent DNA fragments to complete the process. This completes the replication process in the lagging strand.

Replication is Terminated when the Replication Forks Meet at the Terminus Sequences

On the opposite side from *oriC*, the *E. coli* chromosome contains a pair of *termination sequences* called *ter* sequences. A protein known as the *ter*-binding protein binds to the *ter* sequences and stops the movement of the replication forks. One of the *ter* sequences stops the clockwise-moving fork while the other *ter* sequence stops the counterclockwise-moving fork. Therefore, DNA replication ends at the *terminus* sequences and DNA ligase covalently links all four DNA strands, creating two circular double-stranded molecules. Curiously, the termination sequences are arranged so that the forks would pass each other to reach the termination sequence that stops its own advancement. Since this could not occur, one explanation is that only one *ter* sequence is required to stop the advancement of one replication fork, and that the other fork ends its synthesis of DNA when it reaches the halted replication fork.

After DNA replication is completed, one last problem may exist. Bacterial DNA replication often results in two intertwined DNA molecules. Interlocked circular molecules are known as *catenanes*.

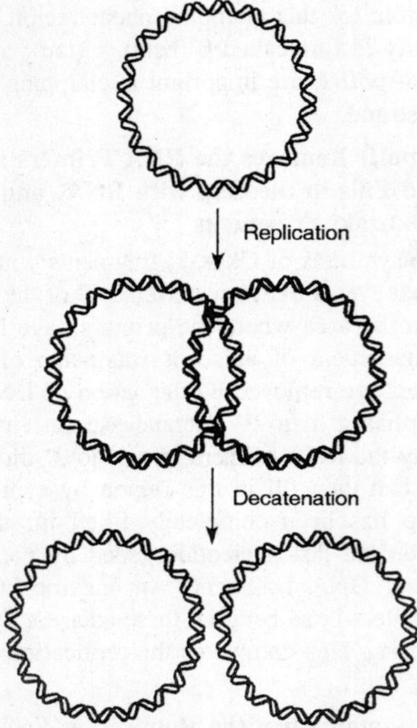

Fig. 2.5. Separation of catenanes.

Fortunately, catenanes are only transient structures in DNA replication. Topoisomerases introduce a temporary break into the DNA strands and then rejoin them after the strands have become unlocked. This allows the catenanes to be separated into individual circular molecules.

Certain Enzymes of DNA Replication Bind to Each Other to Form a Complex

DNA helicase, primase, and several necessary proteins form a multiprotein complex known as a *primosome*. This complex leads the way at the replication fork. The primosome tracks along the DNA processively, separating the parental strands and synthesizing RNA primers at regular intervals along the lagging strand. By acting within a primosome, the actions of helicase and primase can be better coordinated.

Behind the primosome are the other proteins involved in DNA replication. In this drawing, it is suggested that two DNA polymerases (*polIII*) act in concert to replicate the leading and lagging strands. In other words, two DNA polymerases associate together to form a dimeric replicative DNA polymerase that moves as a unit. For this to occur, the lagging strand is looped around one of the DNA polymerase enzymes so that it can synthesize DNA in a 5' to 3' direction, yet move toward the opening of the replication fork. Research also suggests that the DNA polymerases may be associated with the primosome, so that DNA helicase, primase, and the DNA polymerases would move as a single unit along the replication fork. This putative complex has been given the name *replisome*.

The Fidelity of DNA Replication is Ensured by Proofreading Mechanism

With replication occurring so rapidly, one might imagine that mistakes could happen in which the wrong nucleotide could be in corporated into the growing daughter strand. Although mistakes can happen during DNA replication, they are extraordinarily rare. In the case of DNA synthesis via *polIII*, only one mistake per 100,000,000 nucleotides is made. Therefore, DNA synthesis occurs with relatively few errors. Another way of saying this is that DNA replication exhibits a high degree of *fidelity*.

There are several reasons that fidelity is high. First, the hydrogen bonding between G/C and A/T is much more stable than that between mismatched pairs. However, this accounts for only part of the fidelity, since mismatching due to stability considerations would account for one mistake per 1000 nucleotides. Two additional characteristics of DNA polymerases contribute to their fidelity of DNA replication. The active site of DNA polymerase is such that it will preferentially catalyze the attachment of nucleotides when the correct bases are hydrogen bonding in opposite strands. In other words, DNA polymerase is unlikely to catalyze bond formation between adjacent nucleotides if a mismatched base pair is formed. This induced-fit phenomenon decreases the error rate to one in 100,000 to 1,000,000.

Another way that DNA polymerase decreases the error rate is by the enzymatic removal of mismatched nucleotides. DNA polymerase can identify a mismatched nucleotide and remove it from the daughter strand. This occurs by cleavage of the bond between adjacent nucleotides at the 3' end of the newly made strand. This process is referred to as *exonuclease* cleavage (the prefix exonucleas indicates that DNA polymerase digests away nucleotides at the end of a DNA strand). The ability to remove mismatched bases by this mechanism has been called the *proofreading function* of DNA polymerase.

Prokaryotic DNA Replication is Coordinated with Cell Division

Bacterial cells can divide into two daughter cells at an amazing rate. Under optimal conditions, certain bacteria such as *E. coli* can divide every 20 to 30 minutes. It is important that DNA replication only take place when a cell is about to divide. If DNA replication occurs too frequently, then there will be too many copies of the bacterial chromosome per cell. Alternatively, if DNA replication does not occur frequently enough, a daughter cell will be left without a chromosome. Therefore, cell division in bacterial cells must coordinate with DNA replication.

Bacterial cells regulate the DNA replication process by controlling the initiation of replication at the origin. This control has been extensively studied in *E. coli*. In this bacterium, several different mechanisms have been proposed to account for the control of DNA replication. In general, the regulation aims to prevent the premature initiation of DNA replication at *oriC*. Two different mechanisms are described here.

First, the initiation of replication is coupled to cell division by the action of the *DnaA* protein. As discussed previously in this chapter, the DnaA protein binds to dnaA boxes within the origin of replication and opens the DNA helix at the A/T-rich region within the origin. To initiate DNA replication, the concentration of the DnaA protein must become high enough to bind all the dnaA boxes. Immediately following DNA replication, there are twice as many dnaA boxes, and so there is insufficient DnaA protein to reinitiate a second round of replication. Also, some of the DnaA protein may be unavailable for DNA binding because it becomes attached to the cell membrane during cell division. Since it takes time to accumulate newly made DnaA protein, DNA replication cannot occur until the daughter cells have had time to grow.

Another way to regulate DNA replication involves the *GATC sites* within *oriC*. These sites can be methylated by an enzyme known as *dam* methylase. The dam methylase recognizes the GATC sequence, binds there, and attaches a methyl group onto the adenine base DNA methylation within *oriC* helps regulate the replication process. Prior to DNA replication, these sites are methylated in both strands. This full methylation of the GATC sites is necessary for DNA replication to be initiated at the origin. Following DNA replication, the newly made strands are not methylated, since adenine rather than methyl adenine is found in the daughter strands The hemimethylated (i.e., half-methylated) DNA cannot initiate replication until after it has become fully methylated. Since it takes several minutes for the *dam* methylase to methylate all the GATC sequences within this region, DNA replication will not occur again too quickly.

Arthur Kornberg's Work Provided a Way to Measure DNA Replication *in vitro*

Much of our understanding of the process of prokaryotic DNA replication has come from thousands of experiments in which DNA replication has been studied *in vitro*. A common experimental strategy is to purify proteins from cell extracts and to determine their roles in the replication process. In other words, purified proteins, nucleotides, template DNA, and so forth can be mixed together in a test tube, and the synthesis of new DNA strands will occur. While at the Washington University and later at Stanford, this approach was pioneered by Arthur Kornberg and his colleagues.

Although we will not discuss the procedures for purifying replication proteins, experiment illustrates an early success in measuring DNA replication *in vitro*. In 1958, not long after the discovery of the double helix, Kornberg and colleagues set out to develop methods for measuring DNA synthesis. They had to base their initial work on hypotheses about the chemical structure of the precursor nucleotides that are involved in DNA replication. Kornberg correctly hypothesized that nucleoside triphosphates are the precursors for DNA synthesis. Also, he knew that nucleoside triphosphates are soluble in an acidic solution, whereas long strands of DNA are not. Rather, DNA strands will precipitate out of solution at an acidic pH. This precipitation event provides a method to separate nucleotides (e.g., flucleoside triphosphates) from strands of DNA.

Hypothesis

DNA synthesis can occur *in vitro* if all the necessary components are present.

Testing hypothesis

Starting material: An extract of proteins from *E. coli*.

1. Mix together the extract of *E. coli* proteins, template DNA that is not radiolabeled, and ^{32}P-radiolabeled deoxynucleoside triphosphates. This is expected to be a complete system that contains everything necessary for DNA synthesis. As a control, a second sample is made in which the template DNA was omitted from the mixture.

2. Incubate the mixture for 30 minutes at 3 7°C.
3. Add perchioric acid to precipitate DNA. It does not precipitate free nucleotides.
4. Centrifuge the tube. (10,000 × g, 3 minutes) Note: The radiolabeled nucleoside triphosphates that have not been incorporated into DNA will remain in the supernatant.
5. Collect the pellet, which contains precipitated DNA and proteins. (The control pellet is not expected to contain DNA.).
6. Count the amount of radioactivity in the pellet using a scintillation counter.

Data

Conditions	*Amount of Radiolabeled DNA*
Complete system	3300
Template DNA omitted	0.0

Interpreting the data

When the *E. coli* proteins were mixed with nonlabeled template DNA and radio- labeled deoxynucleoside triphosphates, an acid-precipitable and radiolabeled product was formed. This product must be newly synthesized DNA strands rather than individual nucleotides, since the nucleotides are not acid precipitable. As a control, if nonlabeled template DNA was omitted from the assay, no radiolabeled DNA was made. This is the expected result, since the template DNA is necessary to make new daughter strands. Taken together, these results indicate that this technique can be used to measure the synthesis of DNA *in vitro*.

In the late 1950s, the *in vitro* approach provided an exciting starting point to unravel the complexities of the DNA replication process. Since that time, researchers have been able to purify the individual proteins. This approach still continues, particularly as we try to understand the added complexities of eukaryotic DNA replication.

Eukaryotic DNA Replication

Eukaryotic DNA replication is not as well understood as prokaryotic replication. Much research has been carried out in a variety of experimental organisms, particularly yeast and mammals. Many of these studies have found extensive similarities between the general features of DNA replication in prokaryotes and eukaryotes. For example, the types of prokaryotic enzymes have also been identified in eukaryotes. DNA helicases, primases, DNA polymerases, single-strand binding proteins, DNA ligases, and topoisomerases have all been found in eukaryotic species. Nevertheless, at the molecular level, eukaryotic DNA replication appears to be substantially more complex. These complexities are related to several features of eukaryotic cells. In particular, eukaryotic cells have larger, linear chromosomes; the chromatin is more tightly packed within nucleosomes; and cell cycle regulation is much more complicated. This section will emphasize some of the unique features of eukaryotic DNA replication.

Initiation Occurs at Multiple Origins of Replication on Linear Eukaryotic Chromosomes

Since eukaryotes have long linear chromosomes, they require multiple origins of replication so that the DNA can be replicated in a reasonable length of time. DNA replication proceeds bidirectionally from many origins of replications. The multiple replication forks eventually make contact with each other to complete the replication process.

The molecular features of eukaryotic origins of replication may have some similarities to the origins found in bacteria. At the molecular level, eukaryotic origins have been the most extensively studied in the yeast *Saccharomyces cerevisiae*. In this organism, several replication origins have been identified and sequenced. They have been named

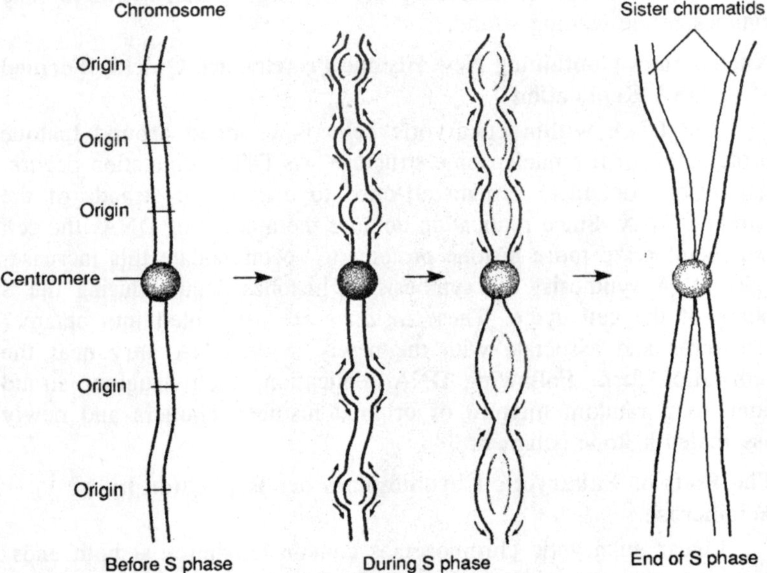

Fig. 2.6. The replication of eukaryotic chromosomes.

ARS elements (autonomously replicating sequence). ARS elements are necessary to initiate chromosome replication *in vivo*. There are two common sequence features among different ARS elements. First, they have a higher percentage of A and T bases than the rest of the chromosomal DNA. Second, the sequence (A or T)TTTAT(A or G)TTT(A or T) is found in all ARS elements. Similar to the dnaA boxes found in bacteria, this sequence may function as a recognition site for an initiator protein. Consistent with this idea is the observation that a complex of six proteins, termed *ORC* (origin recognition complex), appears to bind to this region of the ARS element. Current research is underway to elucidate the mechanism of action of ORC.

Eukaryotes Contain Several Different DNA Polymerases

In mammalian cells, there are at least five different DNA polymerases. Four of these, designated α (alpha), β (beta), δ (delta), and ε (epsilon), function within the cell nucleus, whereas γ (gamma) functions in the mitochondria to replicate mitochondrial DNA. In the nucleus, DNA polymerases β and ε play a role in repairing damaged DNA. DNA polymerases a and are responsible for chromosomal DNA replication. DNA polymerase a contains a subunit that functions as a primase, whereas lacks any primase activity. DNA polymerase a may be responsible for synthesizing the lagging strand, while e may synthesize the leading strand.

Nucleosomes Containing New Histone Proteins are Quickly Formed after DNA Replication

The DNA within eukaryotic cells is wrapped around histone proteins to form a nucleosome structure. As DNA replication occurs, the histone octamers remain attached to one of the strands of the parental DNA. Since replication doubles the amount of DNA, the cell must synthesize more histone proteins to accommodate this increase. Like DNA synthesis, the synthesis of histones occurs during the S phase of the cell cycle. These histones are assembled into octamer structures and associate with the newly made DNA very near the replication fork. Following DNA replication, each daughter strand contains a random mixture of original histone octamers and newly assembled histone octamers.

The Ends of Eukaryotic Chromosomes are Replicated by a Telomerase

Linear eukaryotic chromosomes contain telomeres at both ends. The term *telomere* refers to the complex of telomeric sequences within the DNA and the special proteins that are bound to these sequences. Telomeric sequences consist of a moderately repetitive tandem array and a 3' overhang region that is 12 to 16 nucleotides in length.

The tandem array that occurs within the telomere has been studied in a wide variety of eukaryotic organisms. A common feature is that the telomeric sequence contains several guanine nucleotides and often many thymine nucleotides. Depending on the species and the cell type, this sequence can be tandemly repeated between several and several hundred times in the telomere region.

Telomeres have a specialized form of DNA replication. As discussed previously in this chapter, DNA polymerases only synthesize DNA in a 5' to 3' direction, and they cannot start DNA synthesis at the first two nucleotides (a primer is required here). These two features of DNA polymerase function pose a problem at the ends of linear chromosomes. In particular, one of the strands can not be replicated at the very end by DNA polymerase. Upstream from this point, there is no place on the parental DNA strand for the primer to bind. Therefore, if this problem were not solved, the linear chromosome would become progressively shorter with each round of DNA replication.

Table 2.1. Telomeric sequences within selected organisms

Organism	*Telomeric repeat sequence*
Mammals	AGGGTT
Slime molds	AGGGTT
Filamentous fungi	AGGGTT
Tetrahymena	GGGGTT
Paramecium	GGG[GT]TT
Higher plants	AGGGTTT
Baker's yeast	G(l—3)T

To prevent the loss of genetic information due to Chromosome shortening, eukaryotic cells attach additional DNA sequences to the ends of telomeres. This specialized mechanism involves the attachment of telomeric sequences using an enzyme known as *telomerase*. This enzyme recognizes telomeric sequences at the ends of eukaryotic chromosomes and synthesizes additional numbers of telomeric repeat sequences. The telomerase enzyme contains both protein and RNA. The RNA part of telomerase contains a sequence that is complementary to the DNA sequence found in the telomeric repeat. This allows telomerase to bind to the end of the telomere. Following binding, the RNA sequence beyond the binding site can also function as a template allowing the attachment of a six-nucleotide sequence to the end of the DNA strand. This is called *polymerization*, since it is analogous to the function of DNA polymerase. The telomerase can then move (i.e., translocate) to the new end of this DNA strand and attach another six nucleotides to the end. And so on. This binding—polymerization—translocation cycle can occur many times and thereby greatly lengthen one of the DNA strands in the telomeric region. The complementary strand is synthesized by DNA polymerase. In this way, the progressive shortening of eukaryotic chromosomes is prevented.

Concluding Remark

The structural basis for *DNA replication* is the double-stranded helix of DNA, in which the A—TIG—C rule is obeyed. This complementarity between strands allows DNA to be replicated in a *semiconservative* fashion. DNA replication begins at a specific site within the DNA, known as the origin of replication. Circular bacterial chromosomes have a single origin of replication. Specific proteins, such as the DnaA protein found in *E. coli*, recognize the origin and initiate the DNA replication process. The synthesis of DNA strands proceeds *bidirectionally* from the origin. This synthesis requires various proteins. *Helicase* breaks the hydrogen bonding between the *parental strands*, *topoisomerase* removes positive supercoils, and *single-strand binding proteins* hold the parental strands in a single-stranded state. *Primase* synthesizes RNA primers, which are necessary for *DNA polymerase* to elongate new *daughter strands* in a 5' to 3' direction. In the *leading strand*, synthesis of the daughter strand is continuous, In the lagging strand, the synthesis of DNA occurs in short *Okazaki fragments*. The RNA primers are removed, DNA polymerase fills in the gaps, and DNA ligase covalently attaches the fragments together. DNA replication is terminated when both replication forks reach the ter sequences, and a possible catenane structure is resolved.

Coordination of cell division and DNA replication is accomplished via the cellular regulation of replication. In bacteria, the control of DNA replication occurs at the origin. Several mechanisms can prevent premature DNA replication. Two examples are the availability of the Dna A protein and the methylation of GATC sites.

DNA replication in eukaryotes has several unique features. For example, eukaryotic chromosomes contain multiple origins of replication. Also, since eukaryotic chromosomes are linear, a specialized mechanism exists for the replication of the ends of the chromosomes within the telomere. An enzyme known as telomerase at taches telomeric repeat sequences. This prevents chromosome shortening with each round of DNA replication. Also, since eukaryotic chromosomes exist in a nucleosome structure, it is necessary to assemble the DNA and histone octamers following replication. This assembly occurs in the vicinity of the replication fork, and replicated chromosomes contain a random mixture of new and old histone octamers.

The semiconservative mechanism of DNA replication was initially proposed based on the double helix structure determined by Watson and Crick. Nevertheless, researchers needed to experimentally demonstrate that this mechanism was correct. The isotope labeling experiments of Meselson and Stahl were consistent with a semi-conservative mode of replication and were inconsistent with conservative and dispersive models.

Once the semiconservative model was established, researchers focused their attention on the details of DNA replication within living cells. Arthur Kornberg and his colleagues developed methods to synthesize bacterial DNA *in vitro*. This allowed researchers to identify the components required for DNA replication. Many of these components were found to be enzymes, such as DNA polymerase, primase, helicase, topoisomerase, and ligase. Also, an analysis of DNA revealed two sequences, *oriC* and *ter*, that are necessary for the initiation and termination of DNA replication. In eukaryotes, similar findings have been obtained, but there are more enzymes and the sequences of the origins of replications are more complex. These added complexities have made it more difficult to analyze DNA replication in eukaryotes, although much progress has been made.

3

DNA—Repair and Recombination

The mutation, repair, and recombination of DNA are treated together in this chapter because the three processes have much in common. The physical alteration of DNA is involved in each; repair and recombination share some of the same enzymes. We progress from mutation —the change in DNA—to repair of damaged DNA, and, finally, to recombination, the new arrangement of pieces of DNA.

Mutation

The concept of mutation (a term coined by de Vries, a rediscoverer of Mendel) is pervasive in genetics. *Mutation* is both the process by which a gene (or chromosome) changes structurally and the end result of that process. Without alternative forms of genes, the biological diversity that exists today could not have evolved. Without alternative forms of genes, it would have been virtually impossible for geneticists to determine which of an organism's characteristics are genetically controlled. Studies of mutation provided the background for our current knowledge in genetics.

Fluctuation Test

In 1943, Salvador Luria and Max Delbrück published a paper entitled "Mutations of Bacteria from Virus Sensitivity to Virus Resistance." This paper ushered in the era of bacterial genetics by demonstrating that the phenotypic variants found in bacteria are actually attributable to mutations rather than to induced physiological changes. Very little work had previously been done in bacterial genetics because of the feeling that bacteria did not have "normal" genetic systems like the systems of fruit flies and corn. Rather, bacteria were believed to respond to environmental change by physiological adaptation, a non-Darwinian view. As Luria said, bacteriology remained "the last stronghold of Lamarckism" (the belief that acquired characteristics are inherited).

What causes genetic variation?

Luria and Delbrück studied the Ton^r (phage T1-resistant) mutants of a normal Ton^s (phage T1-sensitive) *Escherichia coli* strain. They used an enrichment experiment, wherein a petri plate is spread with *E. coli* bacteria and T1 phages. Normally, no bacterial colonies grow on the plate: all the bacteria are lysed. However, if one of the bacterial cells is resistant to T1 phages, it produces a bacterial colony, and all descendants of this colony are T1 resistant. There are two possible explanations for the appearance of T1-resistant colonies:

1. Any *E. coli* cell may be induced to be resistant to phage T1, but only a very small number actually are. That is, all cells are genetically identical, each with a very low probability of exhibiting resistance in the presence of T1 phages. When resistance is induced, the cell and its progeny remain resistant.
2. In the culture, a small number of *E. coli* cells exist that are already resistant to phage T1; in the presence of phage T1, only these cells survive.

If the presumed rates of physiological induction and mutation are the same, determining which of the two mechanisms is operating is difficult. Luria and Delbrück, however, developed a means of distinguishing between these mechanisms. They reasoned as follows: If T1 resistance was physiologically induced, the relative frequency of resistant *E. coli* cells in a culture of the normal (Ton^s) strain should be a constant, independent of the number of cells in the culture or the length of time that the culture has been growing. If resistance was due to random mutation, the frequency of mutant (Ton^r) cells would depend on when the mutations occurred. In other words, the appearance of a mutant cell would be a random event. If a mutation occurs early in the growth of the culture, then many cells descend from the mutant cell, and therefore many resistant colonies develop. If the mutation does not occur until late in the growth of the culture, then the subsequent number of mutant cells is few. Thus, if the mutation hypothesis is correct, there should be considerable fluctuation from culture to culture in the number of resistant cells present.

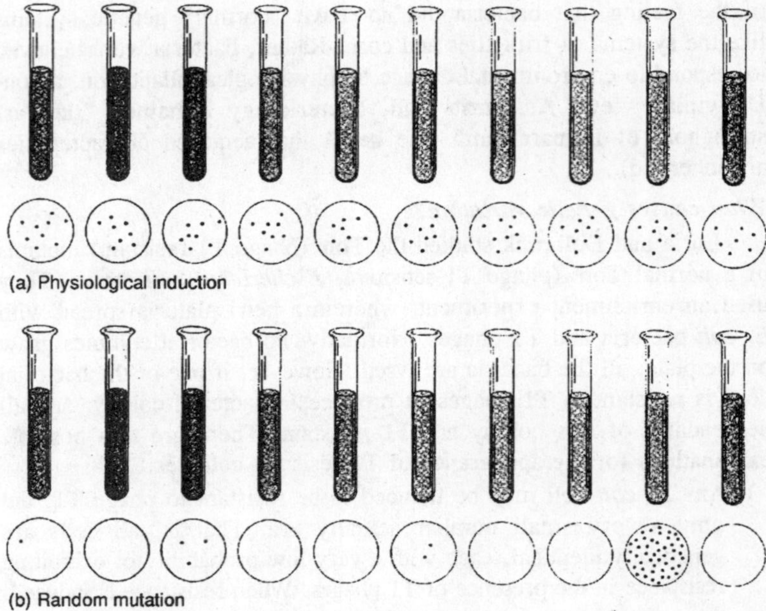

Fig. 3.1. Occurrence of E. coli Ton^r colonies in Ton^s cultures.

Results of the fluctuation test

To distinguish between these hypotheses, Luria and Delbrück developed what is known as the *fluctuation test*. They counted the mutants both in small ("individual") cultures and in sub-samples from a single large ("bulk") culture. All sub-samples from a bulk culture should have the same number of resistant cells, differing only because of random sampling error. If, however, mutation occurs, the number of resistant cells among the individual cultures should vary considerably from culture to culture; the number would be related to the time that the mutation occurred during the growth of each culture. If mutation arose early, there would be many resistant cells. If it arose late, there would be relatively few resistant cells. Under physiological induction, the distribution of resistant colonies should not differ between the individual and bulk cultures.

Luria and Delbrück inoculated twenty individual cultures and one bulk culture with *E. coli* cells and incubated them in the absence of phage T1. Each individual culture was then spread out on a petri plate containing a very high concentration of T1 phages; ten sub-samples from the bulk culture were plated in the same way. We can see from the results that there was minimal variation in the number of resistant cells among the bulk culture sub-samples but a very large amount of variation, as predicted for random mutation, among the individual cultures.

If bacteria have "normal" genetic systems that undergo mutation, bacteria could then be used, along with higher organisms, to answer genetic questions. As we have pointed out, the modern era of molecular genetics began with the use of prokaryotic and viral systems in genetic research. In the next section, we turn our attention to several basic questions about the gene, questions whose answers were found in several instances only because prokaryotic systems were available.

Genetic Fine Structure

How do we determine the relationship among several mutations that cause the same phenotypic change? What are the smallest units of DNA capable of mutation and recombination? Are the gene and its protein product colinear? The answers to the latter two questions are important from a historical perspective. The answer to the first question is relevant to our current understanding of genetics.

Complementation

If two recessive mutations arise independently and both have the same phenotype, how do we know whether they are both mutations of the same gene? That is, how do we know whether they are alleles? To answer this question, we must construct a heterozygote and determine the *complementation* between the two mutations. A heterozygote with two mutations of the same gene will produce only mutant messenger RNAs, which result in mutant enzymes. If, however, the mutations are not allelic, the gamete from the a_1 parent will also contain an a_2^+ allele, and the gamete from the a_2 parent will also contain the a_1^+ allele. If the two mutant genes are truly alleles, then the phenotype of the heterozygote should be mutant. If, however, the two mutant genes are nonallelic, then the a_1 mutant will have contributed the wild-type allele at the A_2 locus, and the a_2 mutant will have contributed the wild-type allele at the A_1 locus to the heterozygote. Thus, the two mutations will complement each other and produce the wild-type. Mutations that fail to complement each other are termed *functional alleles*. The test for defining alleles strictly on this basis of functionality is termed the *cis-trans complementation test*.

There are two different configurations in which a heterozygous double mutant of functional alleles can form. In the *cis-trans* complementation test, only the *trans* configuration is used to determine whether the two mutations were allelic. In reality, the *cis* configuration is not tested; it is the conceptual control, in which wild-type activity (with recessive mutations) is always expected. The test is thus sometimes simply called a *trans* test. Functional alleles produce a wild-type phenotype in the *cis* configuration but a mutant phenotype in the *trans* configuration. This difference in phenotypes is called a *cis-trans* position effect.

From the terms *cis* and *trans,* Seymour Benzer coined the term *cistron* for the smallest genetic unit (length of genetic material) that exhibits a *cis-trans* position effect. We thus have a new word for the gene, one in which function is more explicit. We have, in essence, refined Beadle and Tatum's one-gene-one-enzyme hypothesis to a more accurate one-cistron-one-polypeptide concept. The cistron is the smallest unit that codes for a messenger RNA that is then translated into a single polypeptide or expressed directly (transfer RNA or ribosomal RNA).

From functional alleles, we can go one step further in recombinational analysis by determining whether two allelic mutations occur at exactly the same place in the cistron. In other words, when two mutations prove to be functional alleles, are they also *structural alleles*? The methods used to analyze complementation can be used here also. Crosses are carried out to form a mutant heterozygote (*trans* configuration) whose offspring are then tested for recombination between the two mutational sites. If no recombination occurs, then the two alleles probably contain the same structural change (involving the same base pairs) and are thus structural alleles. If a small amount of recombination occurs that generates wild-type offspring, then the two alleles are not mutations at the same point. Alleles that were functional but not structural were first termed *pseudoalleles* because it was believed that loci were made up of subloci. Fine-structure analysis led to the understanding that a locus is a length of genetic material divisible by recombination rather than a "bead on a string."

Eye-color mutants of *Drosophila melanogaster* can be studied by complementational analysis. The white-eye locus has a series of alleles producing varying shades of red. This locus is sex linked, at about map position 3.0 on the X chromosome. (Several other eye-color loci on the X chromosome are not relevant to this cross—e.g., prune and ruby.) If an apricot-eyed female is mated with a white-eyed male, the female offspring are all heterozygous and have mutant light-colored eyes. Thus, apricot and white are functional alleles: they do not complement. To determine whether apricot and white are structural alleles, light-eyed females are crossed with white-eyed males, and the offspring are observed for the presence of wild-type or light-eyed males. Though their rate of appearance is less than 0.001%, this is significantly above the background mutation rate. The conclusion is that apricot and white are functional, but not structural, alleles.

Fine-structure mapping

After Beadle and Tatum established in 1941 that a gene controls the production of an enzyme that then controls a step in a biochemical pathway, Benzer used analytical techniques to dissect the fine structure of the gene. Fine-structure mapping means examining the size and number of sites within a gene that are capable of mutation and recombination. In the late 1950s, when biochemical techniques were not yet available for DNA sequencing, Benzer used classical recombinational and mutational techniques with bacterial viruses to provide reasonable estimates on the details of fine structure and to give insight into the nature of the gene. He coined the terms *muton* for the smallest mutable site and *recon* for the smallest unit of recombination. It is now known that both muton and recon are a single base pair.

Before Benzer's work, genes were thought of as beads on a string. The very low rate of recombination between sites within a gene hampered the analysis of mutational sites within a gene by means of recombination. If two mutant genes are functional alleles (involving different sites on the same gene), a distinct probability exists that we will get both mutant sites (and both wild-type sites) on the same chromosome by recombination; but, in view of the very short distances within a gene, this probability is very low. Although it certainly seemed desirable to map sites within the gene, the problem of finding an organism that would allow fine-structure analysis remained until Benzer decided to use phage T4.

r II screening techniques

Benzer used the T4 bacteriophage because of the growth potential of phages, in which a generation takes about an hour and the increase in numbers per generation is about a hundredfold. Actually, any prokaryote or virus should suffice, but Benzer made use of other unique screening properties of the phage that made it possible to recognize one particular mutant in about a billion phages. Benzer used *r*II mutants of T4. These mutants produce large, smooth-edged plaques on *E. coli*, whereas the wild-type produces smaller plaques whose edges are not as smooth.

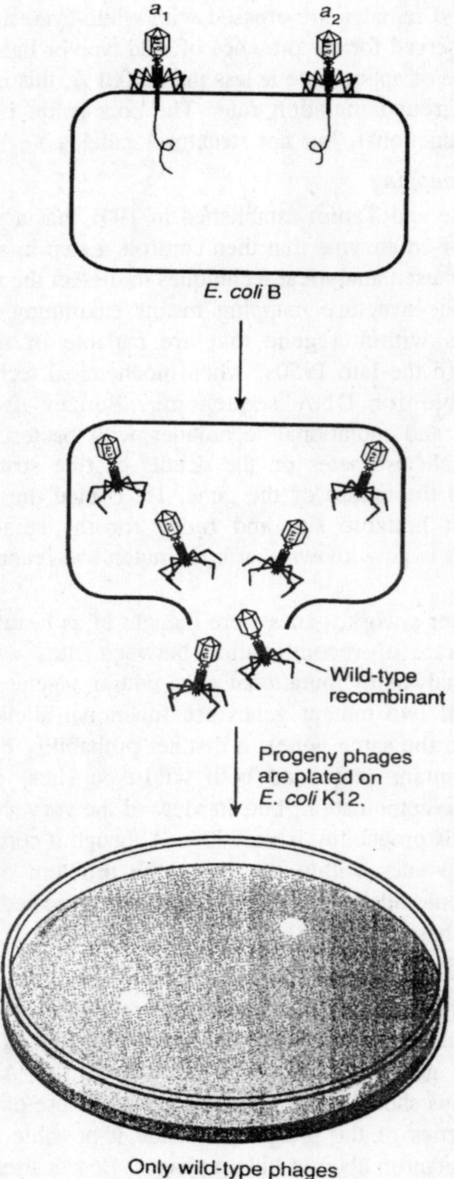

Fig. 3.2. Using E. coli K12 and B strains to screen for recombination at the rII locus of phage T4.

The screening system that Benzer employed made use of the fact that *r*II mutants do not grow on *E. coli* strain K12, whereas the wild-type can. The normal host strain, *E. coli* B, allows growth of both the wild-type and *r*II mutants. Thus, various mutants can be crossed by mixed infection of *E. coli* B cells, and Benzer could screen for wildtype recombinants by plating the resultant progeny phages on *E. coli* K12, on which only a wildtype recombinant produces a plaque. It is possible to detect about one recombinant in a billion phages, all in an afternoon's work. This ability to detect recombinants occurring at such a low level of frequency allowed Benzer to see recombinational events occurring very close together on the DNA, events that would normally occur at a frequency too low to detect in fruit flies or corn.

Benzer sought to map the number of sites subject to recombination and mutation within the *r*II region of T4. He began by isolating independently derived *r*II mutants and crossing them among themselves. The first thing he found was that the *r*II region was composed of two cistrons; almost all of the mutations belonged to one of two *complementation groups*. The *A*-cistron mutations would not complement each other but would complement the mutations of the *B* cistron. The exceptions were mutations that seemed to belong to both cistrons. These mutations were soon found to be deletions in which part of each cistron was missing.

Deletion mapping

As the number of independently isolated mutations of the *A* and *B* cistrons increased, it became obvious that to make every possible pairwise cross would entail millions of crosses. To overcome this problem, Benzer isolated mutants that had partial or complete deletions of each cistron. Deletion mutations were easy to discover because they acted like structural alleles to alleles that were not themselves structurally allelic. In other words, if mutations *a, b,* and *c* are functional—but not structural—alleles of each other, and mutation *d* is a structural allele to *a, b,* and *c,* then *d* must contain a deletion of the bases mutated in *a, b,* and *c.* Once a sequence of deletion mutations covering the *A* and *B* cistrons was isolated, a minimal number of crosses was required to localize a new mutation to a portion of one of the cistrons. A second series of smaller deletions within each region was then isolated, further localizing the mutation.

Next, each new mutant was crossed with each of the other mutants isolated in its subregion to localize the relative position of the new mutation. If the mutation was structurally allelic to a previously isolated mutation, it was scored as an independent isolation of the same mutation. If it was not a structural allele to any of the known mutations of the subregion, it was added as a new mutation point. The exact position of each new mutation within the region was determined by the relative frequency of recombination between it and the known mutations of this region. Benzer eventually isolated about 350 mutations from eighty different subregions defined by deletion mutations.

What conclusions did Benzer draw from his work? First, he concluded that since all of the mutations in both *r*II cistrons can be ordered in a linear fashion, the original Watson-Crick model of DNA as a linear molecule was correct. Second, he concluded that reasonable inroads had been made toward saturating the map, localizing at least one mutation at every mutable site. Benzer reasoned that since many sites were represented by only one mutation, some sites must occur that were represented by zero mutations (i.e., not yet represented by a mutation). Since he had mapped about 350 sites, he calculated that there were at least another 100 sites still undetected by mutation. We now know that 450 sites is an underestimate. However, since the protein products of these cistrons were not isolated, there were no independent estimates of the number of nucleotides in these cistrons (number of amino acids times three nucleotides per codon). Thus, although Benzer had not saturated the map with mutations, he certainly had made respectable progress in dissecting the gene and demonstrating that it was not an indivisible unit, a "bead on a string."

Hot spots

Benzer also looked into the lack of uniformity in the occurrence of mutations. Presuming that all base pairs are either AT or GC, this lack of uniformity was unexpected. Benzer suggested that spontaneous mutation is not just a function of the base pair itself, but is affected by the surrounding bases as well. This concept still holds.

To recapitulate, Benzer's work supports the model of the gene as a linear arrangement of DNA whose nucleotides are the smallest units of mutation. The link between any adjacent nucleotides can break in the recombinational process. The smallest functional unit, determined by a complementation test, is the cistron. Mutagenesis is not uniform throughout the cistron, but may depend on the particular arrangement of bases in a given region.

Intra-allelic complementation

Benzer warned that certainty is elusive in the complementation test because sometimes two mutations of the same functional unit

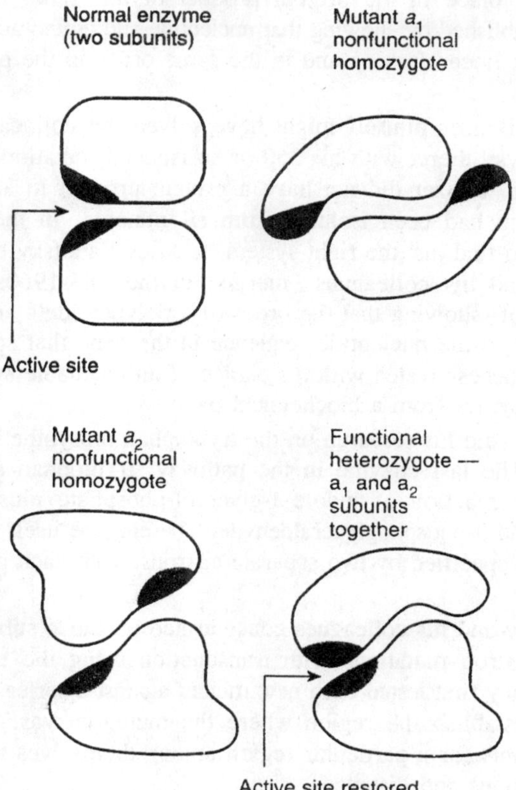

Fig. 3.3. Intra-allelic complementation.

(cistron) can result in partial activity. The problem can be traced to the interactions of subunits at the polypeptide level. Some proteins are made up of subunits, and it is possible that certain mutant combinations produce subunits that interact to restore the enzymatic function of the protein. This phenomenon is known as *intra-allelic complementation*. With this in mind, geneticists routinely use the complementation test to determine functional relationships among mutations.

Colinearity

Next we look at the colinearity of the gene and the polypeptide. Benzer's work established that the gene was a linear entity, as Watson and Crick had proposed. However, Benzer could not demonstrate the colinearity of the gene and its protein product. To do this, it is necessary to show that for every mutational change in the DNA, a corresponding change takes place in the protein product of the gene. Colinearity would be established by showing that nucleotide and amino acid changes occurred in a linear fashion and in the same order in the protein and in the cistron.

Ideally, Benzer himself might have solved the colinearity issue. He was halfway there, with his 350 or so isolated mutations of phage T4. However, Benzer did not have a protein product to analyze; no mutant protein had been isolated from *r*II mutants. In the midst of competition to find just the right system, Charles Yanofsky of Stanford University and his colleagues emerged in the mid-1960s with the required proof, showing that the order of a polypeptide's amino acids corresponded to the nucleotide sequence in the gene that specified it. Yanofsky's success rested with his choice of an amenable system, one using the enzymes from a biochemical pathway.

Yanofsky did his research on the tryptophan biosynthetic pathway in *E. coli*. The last enzyme in the pathway, tryptophan synthetase, catalyzes the reaction of indole-3-glycerol-phosphate plus serine to tryptophan and 3-phosphoglyceraldehyde. The enzyme itself is made of four subunits specified by two separate cistrons, with each polypeptide present twice.

Yanofsky and his colleagues concentrated on the *A* subunit. They mapped *A*-cistron mutations with transduction using the transducing phage P1. They first tested each new mutant against a series of deletion mutants to establish the region where the mutation was. Then they crossed mutants for a particular region among themselves to establish relative positions and distances.

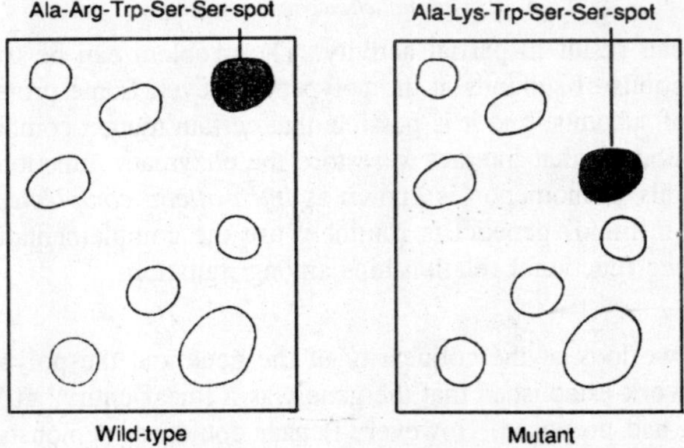

Fig. 3.4. Difference in "fingerprints" between mutant and wild-type polypeptide digests.

The protein products of the bacterial genes were isolated using electrophoresis and chromatography to establish the fingerprint patterns of the proteins. Assuming a single mutation, a comparison of the mutant and the wild-type fingerprints would show a difference of just one polypeptide spot, avoiding the need to sequence the entire protein. The mutant amino acid was identified by analysis of just this one spot.

We can see from this figure that nine mutations in the linear *A* cistron of tryptophan synthetase are colinear with nine amino acid changes in the protein itself. In two cases, two mutations mapped so close as to be almost indistinguishable. In both cases, the two mutations proved to be in the same codon: the same amino acid position was altered in each (A23–A46, A58–A78). Thus, exactly as predicted and expected, colinearity exists between gene and protein. Brenner and his colleagues, using head-protein mutants of phage T4, independently confirmed this work at the same time.

Spontaneous Versus Induced Mutation

H. J. Muller won the Nobel Prize for demonstrating that X rays can cause mutations. This work was published in 1927 in a paper entitled "Artificial Transmutation of the Gene." At about the same time, L. J. Stadler induced mutations in barley with X rays. The basic impetus for their work was the fact that mutations occur so infrequently that genetic research was hampered by the inability to obtain mutants. Muller exposed flies to varying doses of X rays and then observed their progeny. He came to several conclusions. First, X rays greatly increased the occurrence of mutations. Second, the inheritance patterns of X-ray-induced mutations and the resulting phenotypes of organisms were similar to those that resulted from natural, or "spontaneous," mutations.

Mutation Rates

The *mutation rate* is the number of mutations that arise per cell division in bacteria and single-celled organisms, or the number of mutations that arise per gamete in higher organisms. Mutation rates vary tremendously depending upon the length of genetic material, the kind of mutation, and other factors. Luria and Delbrück, for example, found that in *E. coli* the mutation rate per cell division of Ton^s to Ton^r was 3×10^{-8}, whereas the mutation rate of the wild-type to the histidine-requiring phenotype (His^+ to His^-) was 2×10^{-6}. The rate of *reversion* (return of the mutant to the wild-type) was 7.5×10^{-9}. The mutation and reversion rates differ because many different mutations can cause the His phenotype, whereas reversion requires specific, and hence less probable, changes to correct the His phenotype back to the wildtype. The lethal mutation rate in *Drosophila* is about 1×10^{-2} per gamete for the total genome. This number is relatively large because, as with His, many different mutations produce the same phenotype (lethality, in this case).

Point Mutations

The mutations of primary concern in this chapter are *point mutations*, which consist of single changes in the nucleotide sequence. If the change is a replacement of some kind, then a new codon is created. In many cases, this new codon, upon translation, results in a new amino acid. As discussed already that one of the outcomes of redundancy in the genetic code is partial protection of the cell from the effects of mutation; common amino acids have the most codons, similar amino acids have similar codons, and the wobble position of the codon is the least important position in translation. However, when base changes result in new amino acids, new proteins appear. These new proteins can alter the morphology or physiology of the organism and result in phenotypic novelty or lethality.

Frameshift mutation

A point mutation may consist of replacement, addition, or deletion of a base. Point mutations that add or subtract a base are, potentially, the most devastating in their effects on the cell or organism because they change the reading frame of a gene from the site of mutation onward. A frameshift mutation causes two problems. First, all the codons from the frameshift on will be different and thus yield (most probably) a useless protein. Second, stop-signal information will be misread. One of the new codons may be a nonsense codon, which causes translation to stop prematurely. Or, if the translation apparatus reaches the original nonsense codon, it is no longer recognized as such because it is in a different reading frame, and therefore, the translation process continues beyond the end of the gene.

Back mutation and suppression

A second point mutation in the same gene can have one of three possible effects. First, the mutation can result in either another mutant codon or in one codon that has experienced two changes. Second, if the change is at the same site, the original sequence can be returned, an effect known as *back mutation*: the gene then becomes a revertant, with its original function restored. Third, *intragenic suppression* can

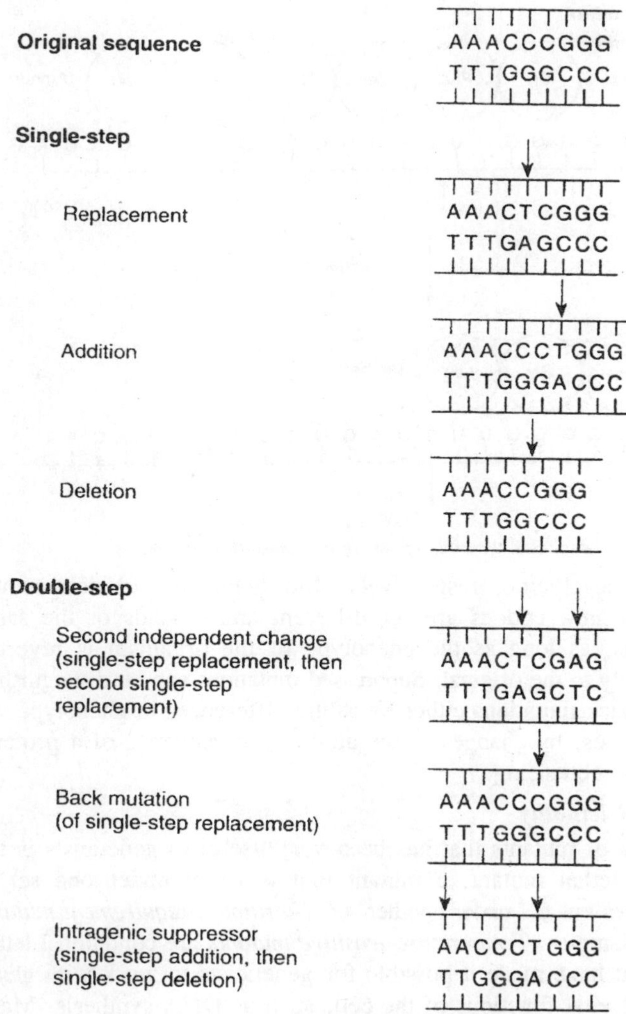

Fig. 3.5. Types of DNA point mutations.

take place. Intragenic suppression occurs when a second mutation in the same gene masks the occurrence of the original mutation without actually restoring the original sequence. The new sequence is a double mutation that appears to have the original (unmutated) phenotype. In figure, a T addition is followed by an A deletion that substitutes the AACCCT sequence for the original AAACCC. These sequences, when transcribed (UUGGGA, UUUGGG), are codons for leucine-glycine and

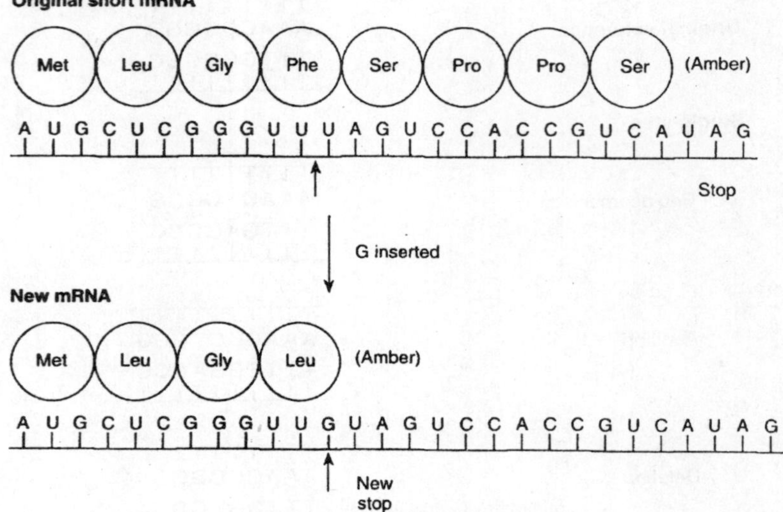

Fig. 3.6. Possible effects of a frameshift mutation.

phenylalanine-glycine, respectively. Intragenic suppression occurs whether the new codons are for different amino acids or the same amino acids, as long as the phenotype of the organism is reverted approximately to the original. Suppressed mutations can be distinguished from true back mutations either by subtle differences in phenotype, by genetic crosses, by changes in the amino acid sequence of a protein, or by DNA sequencing.

Conditional lethality

A class of mutants that has been very useful to geneticists is the conditional-lethal mutant, a mutant that is lethal under one set of circumstances but not under another set. *Nutritional-requirement mutants* are good examples. *Temperature-sensitive mutants* are conditional-lethal mutants that have made it possible for geneticists to work with genes that control vital functions of the cell, such as DNA synthesis. Many temperature-sensitive mutants are completely normal at 25°C but cannot synthesize DNA at 42°C. Presumably, temperature-sensitive mutations result in enzymes with amino acid substitutions that cause protein denaturation to occur at temperatures above normal. Thus, the enzyme has normal function at 25°C, the *permissive temperature*, but is nonfunctional at 42°C, the *restrictive* temperature.

The interesting thing about most conditional-lethal mutants of *E. coli* that cannot synthesize DNA at the restrictive temperature is that they have a completely normal DNA polymerase I. From this information, we infer that polymerase I is not the enzyme *E. coli* normally uses for DNA replication. When an organism with a conditional mutation of polymerase I was isolated, it was able to replicate its DNA normally, but unable to repair damage to the DNA. This led to the conclusion that polymerase I is primarily involved in repair rather than replication of DNA. Conditional-lethal mutants thus allow genetic analysis on genes otherwise impossible to study.

Spontaneous Mutagenesis

Watson and Crick originally suggested that mutation could occur spontaneously during DNA replication if pairing errors occurred. If a

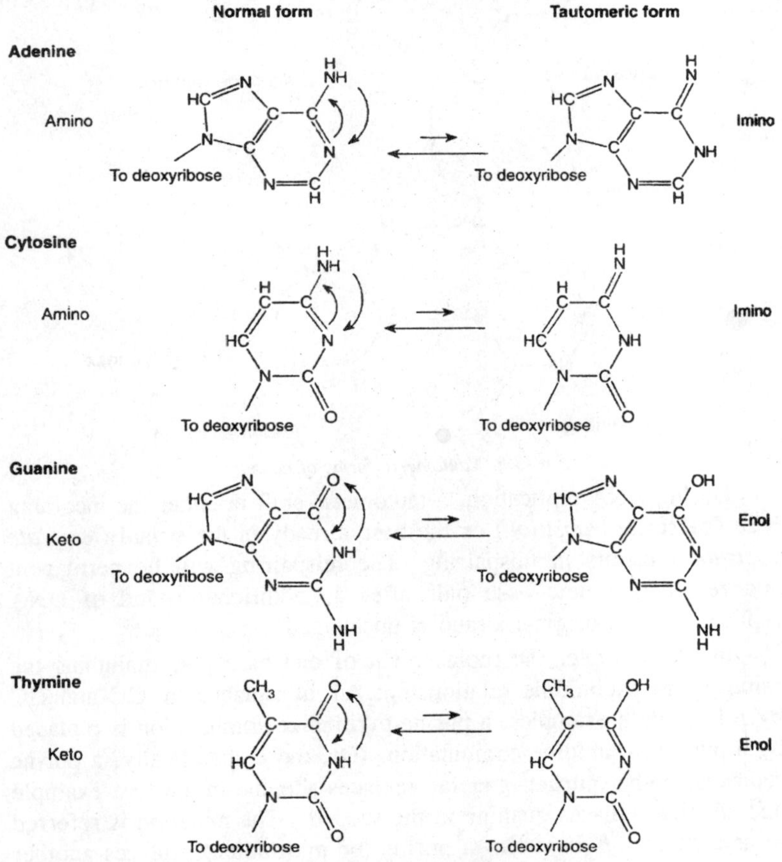

Fig. 3.7. Normal and tautomeric forms of DNA bases.

base of the DNA underwent a proton shift into one of its rare tautomeric forms (*tautomeric shift*) during the replication process, an inappropriate pairing of bases would occur. Normally, adenine and cytosine are in the amino (NH_2) form. Their tautomeric shifts are to the imino (NH) form. Similarly, guanine and thymine go from a keto (C=O) form to an enol (COH) form.

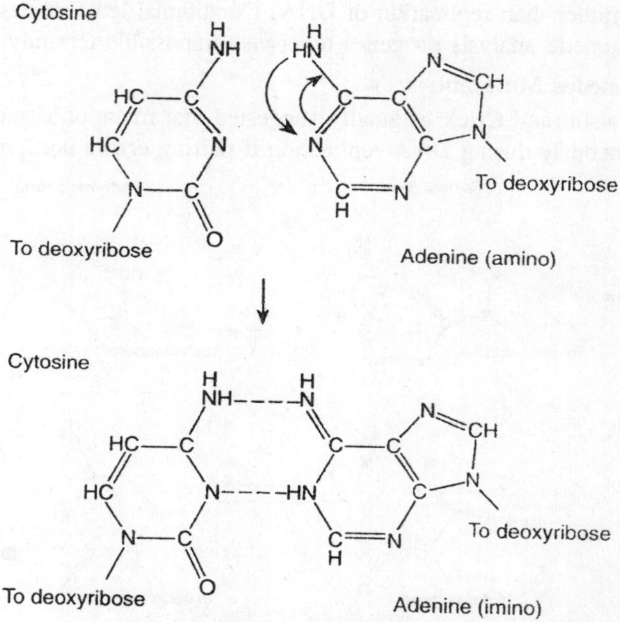

Cytosine

To deoxyribose

Adenine (amino)

To deoxyribose

Cytosine

To deoxyribose

Adenine (imino)

To deoxyribose

Fig. 3.8. Tautomeric forms of adenine.

During DNA replication, a tautomeric shift in either the incoming base (*substrate transition*) or the base already in the strand (*template transition*) results in mispairing. The mispairing will be permanent and result in a new base pair after an additional round of DNA replication. The original strand is unchanged.

In the example, the replacement of one base pair maintains the same purine-pyrimidine relationship: AT is replaced by GC and GC by AT. In both examples, a purine-pyrimidine combination is replaced by a purine-pyrimidine combination. (Or, more specifically, a purine replaces another purine: guanine replaces adenine in the first example and adenine replaces guanine in the second.) The mutation is referred to as a *transition mutation*: a purine (or pyrimidine) replaces another purine (or pyrimidine) through a transitional state involving a tautomeric

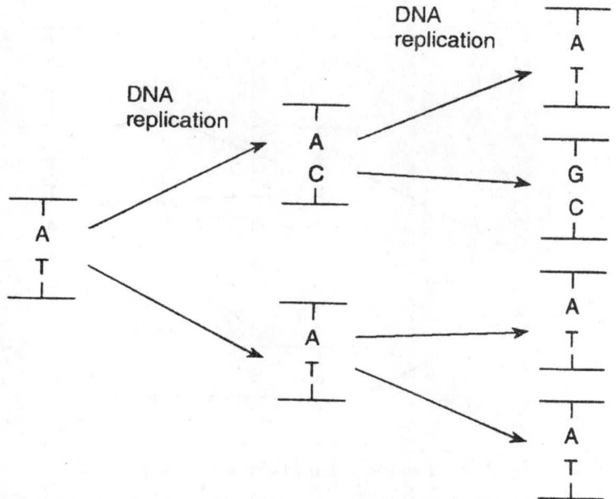

Template transition—tautomerization of adenine in the template

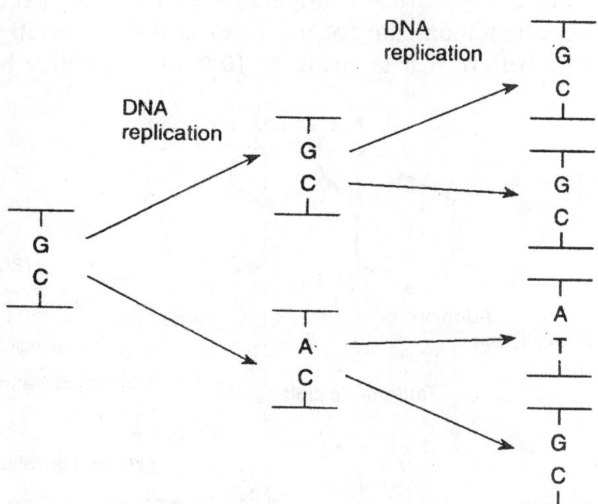

Substrate transition—tautomerization of incoming adenine

Fig. 3.9. Tautomeric shifts result in transition mutations.

shift. When a purine replaces a pyrimidine or vice versa, it is referred to as a ***transversion*** mutation.

Transversions may arise by a combination of two events, a tautomerization and a base rotation. For example, an AT base pair can be converted to a TA base pair (a transversion) by an intermediate

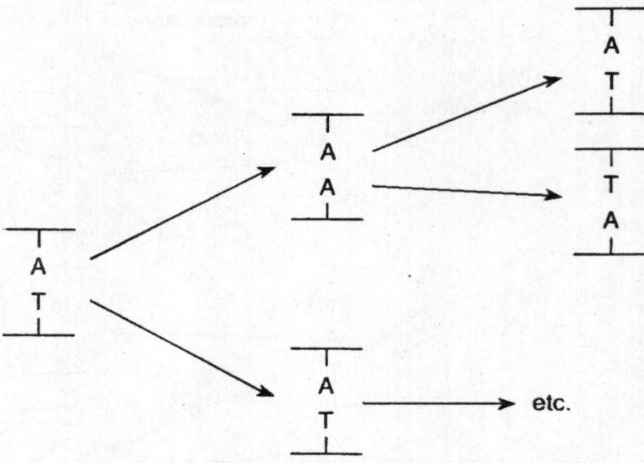

Fig. 3.10. A model for transversion mutagenesis.

AA pairing. Adenine can pair with adenine if one of the bases undergoes a tautomeric shift while the other rotates about its base-sugar (glycosidic) bond. The normal configuration of the base is referred to as the *anti* configuration; the rotated form is the *syn* configuration. Since we now believe that as many as 10% of bases may be in the

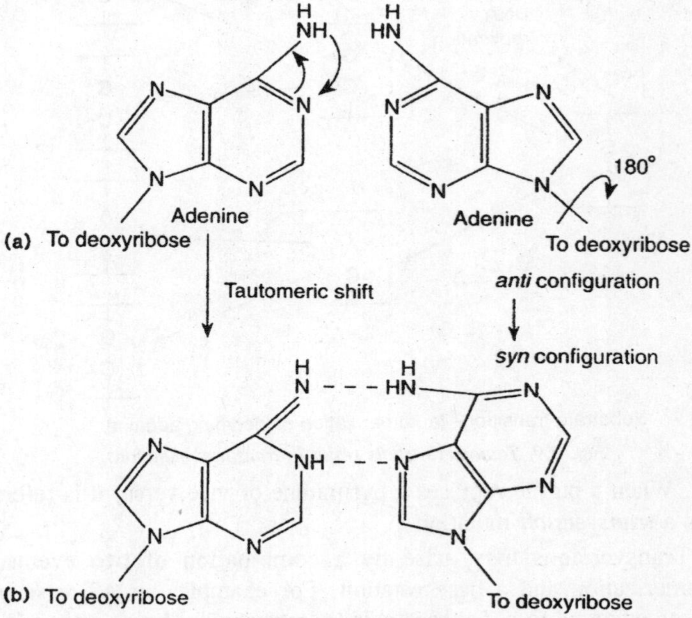

(a) To deoxyribose
Adenine
Adenine
To deoxyribose
180°
Tautomeric shift
anti configuration
syn configuration
(b) To deoxyribose
To deoxyribose

Fig. 3.11. Transversion mutagenesis.

syn configuration at any moment, the transversion mutagenesis rate should be about 10% of the transition mutagenesis rate, a value not inconsistent with current information.

Some base-pair mutations can have serious results. If guanine undergoes an oxidation to 8-oxoguanine, it pairs with adenine. A GC base pair is converted to a TA base pair through an 8-oxoguanine-adenine intermediate. This transversion has been found to be common in cancers.

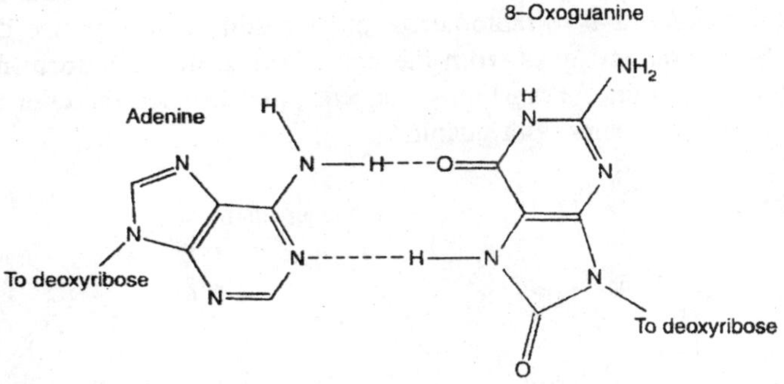

Fig. 3.12. The structure of 8-oxoguanine, which base pairs with adenine, converting a GC base pair to a TA base pair—a transversion.

Since 1953, when Watson and Crick first described the structure of DNA, tautomerization has been accepted as the obvious source of most transition mutations. However, recent structural data has cast some doubt on this assumption. X-ray crystallography and nuclear magnetic resonance (NMR) studies indicate that both bases in transition mismatches may be in their normal forms. Other mechanisms, similar to wobble base pairing, may be responsible for most transition mutations. These studies also indicate that some transversions result from direct purine-purine or pyrimidine-pyrimidine base pairing during DNA synthesis. Much work needs to be done to clarify the nature of spontaneous mutagenesis.

Chemical Mutagenesis

Muller demonstrated that X rays can cause nutation. Certain chemical and temperature treatments can also cause mutation. Determining the mode of action of various chemical mutagens has provided insight into the mutational process as well as the process of carcinogenesis. In addition, knowing how chemical mutagens act has allowed geneticists to produce large numbers of specific mutations at will.

Transitions

Transitions are routinely produced by base analogues. Two of the most widely used base analogues are the pyrimidine analogue 5-bromouracil (5BU) and the purine analogue 2-aminopurine (2AP). The mutagenic mechanisms of the two are similar. The 5-bromouracil is incorporated into DNA in place of thymine; it acts just like thymine in DNA replication and, since it doesn't alter the hydrogen bonding, should induce no mutation. However, it seems that the bromine atom causes 5-bromouracil to tautomerize more readily than thymine does. Thus, 5-bromouracil goes from the keto form to the enol form more readily than thymine. Transitions frequently result when the enol form of 5-bromouracil pairs with guanine.

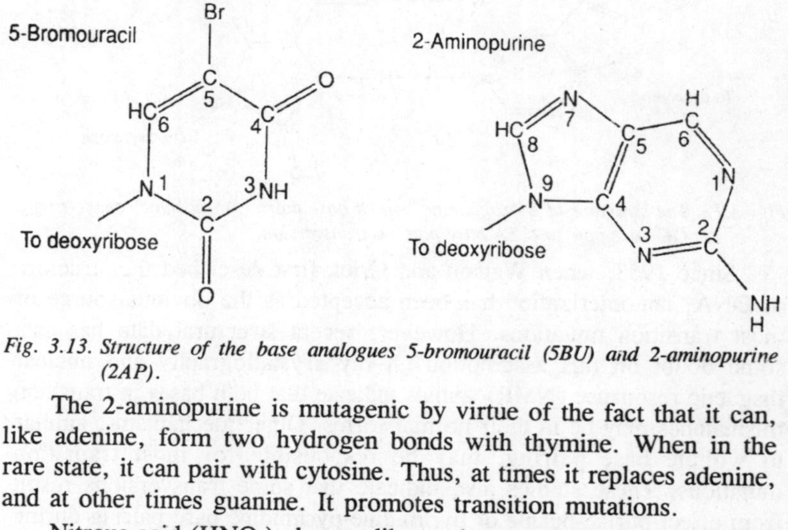

5-Bromouracil
Br
C 5
O
HC 6
4 C
N 1
3 NH
2 C
O
To deoxyribose

2-Aminopurine
N 7
H
C
HC 8
C 5
6
1 N
9 N
C 4
3 N
2 C
NH
H
To deoxyribose

Fig. 3.13. Structure of the base analogues 5-bromouracil (5BU) and 2-aminopurine (2AP).

The 2-aminopurine is mutagenic by virtue of the fact that it can, like adenine, form two hydrogen bonds with thymine. When in the rare state, it can pair with cytosine. Thus, at times it replaces adenine, and at other times guanine. It promotes transition mutations.

Nitrous acid (HNO_2) also readily produces transitions by replacing amino groups on nucleotides with keto groups ($-NH_2$ to $=O$). The result is that cytosine is converted to uracil, adenine to hypoxanthine, and guanine to xanthine. Uracil pairs with adenine instead of guanine, thus leading to a UA base pair in place of a CG base pair; hypoxanthine (H) pairs with cytosine instead of thymine, the original base paired with adenine. Thus, in this case, an HC base pair replaces an AT base pair. Both of these base pairs (UA and HC) are transition mutations. Xanthine, however, pairs with cytosine just as guanine does. Thus, the replacement of guanine with xanthine does not cause changes in base pairing.

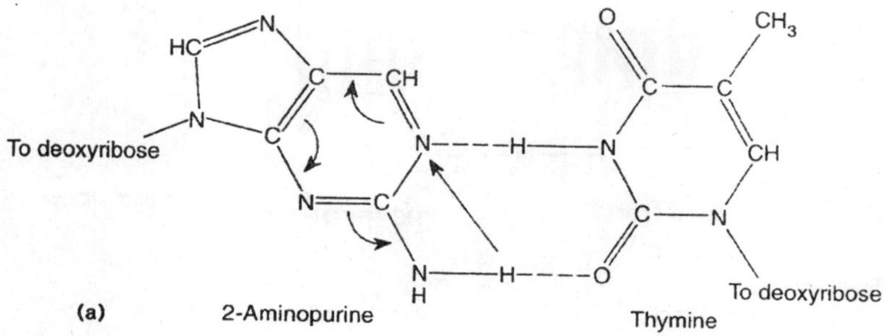

(a) 2-Aminopurine Thymine

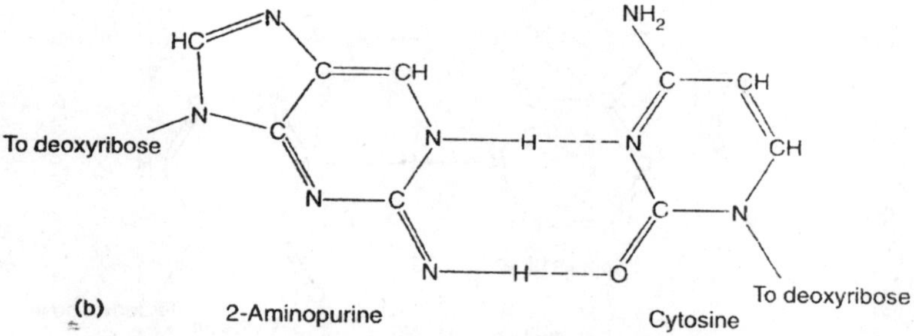

(b) 2-Aminopurine Cytosine

Fig. 3.14. Two possible base pairs with 2-aminopurine (a) in the normal state, 2-aminopurine acts like adenine and pairs with thymine. (b) In the rare state, 2-aminopurine acts like guanine and forms complementary base pairs with cytosine.

Like nitrous acid, heat can also deaminate cytosine to form uracil and thus bring about transitions (CG to TA). Apparently, heat can also bring about transversions by an unknown mechanism.

Transversions

Ethyl methane sulfonate ($CH_3SO_3CH_2CH_3$) and ethyl ethane sulfonate ($CH_3CH_2SO_3CH_2CH_3$) are agents that cause the removal of purine rings from DNA. The multistep process begins with the ethylation of a purine ring and ends with the hydrolysis of the glycosidic (purine-deoxyribose) bond, causing the loss of the base. These sites where this happens are referred to as AP (apurinic-apyrimidinic) sites. If the AP site is not repaired, any of the four DNA bases could be inserted into the new strand opposite the gap. If thymine is placed in the newly formed strand, then the original base pair is restored; insertion of cytosine results in a transition mutation; insertion of either adenine or guanine results in a transversion mutation. Of course, the

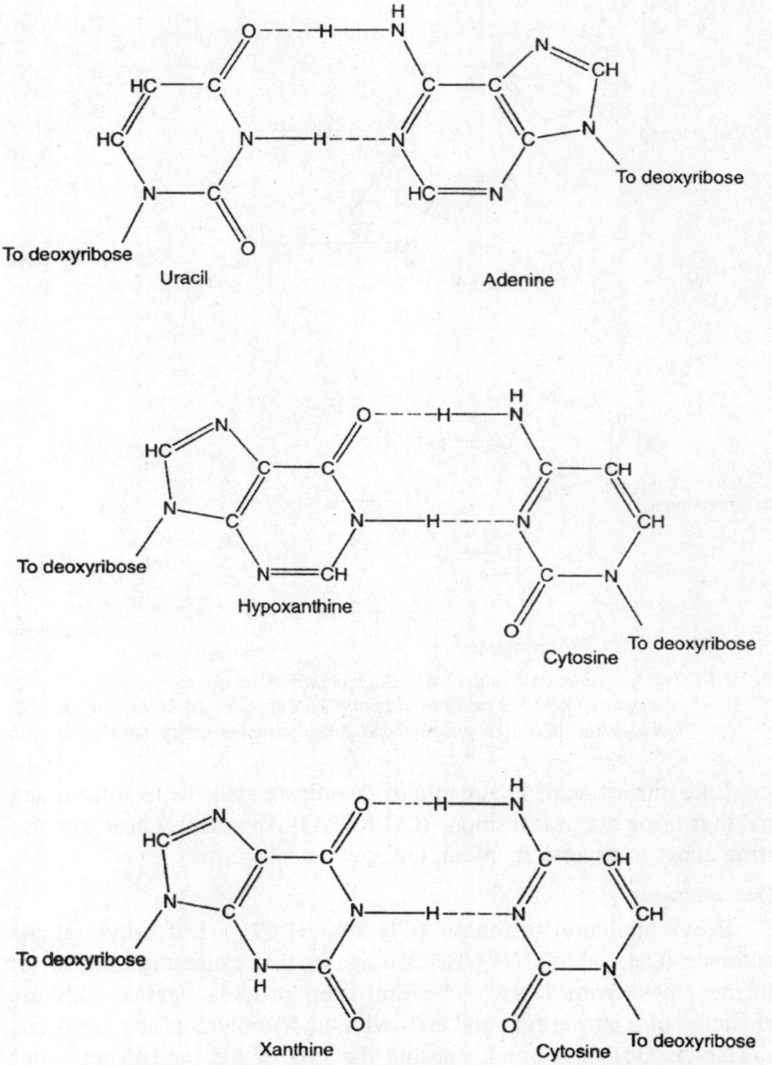

Fig. 3.15. Nitrous acid converts cytosine to uracil, adenine to hypoxanthine, and guanine to xanthine.

gap is still there, and it continues to generate new mutations each generation until it is repaired. During DNA replication in *E. coli*, the polymerase tends to place adenine opposite the gap more frequently than it places other bases.

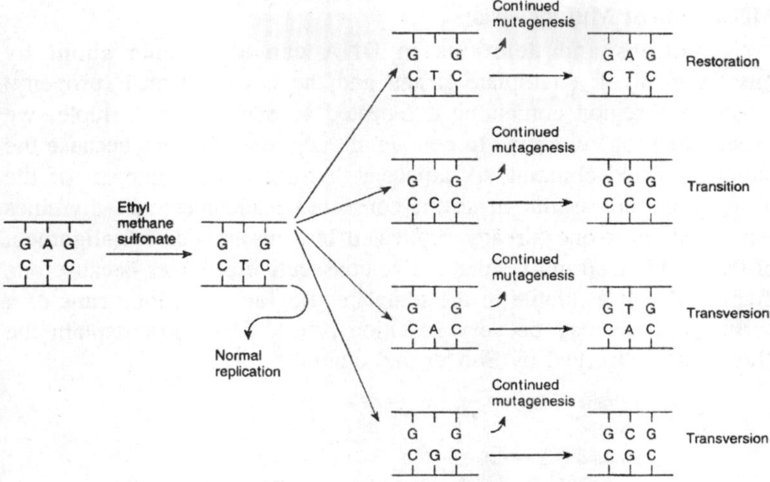

Fig. 3.16. Four possible outcomes after treatment of DNA with an alkylating agent, which removes the purine—adenine in this example.

Insertions and deletions

The molecules of the acridine dyes, such as proflavin and acridine orange, are flat. Presumably, they initiate mutation by inserting into the DNA double helix, causing the helix to buckle in the region of insertion, possibly leading to base additions and deletions during DNA replication. Crick and Brenner used acridine-induced mutations to demonstrate both that the genetic code was read from a fixed point and that it was triplet.

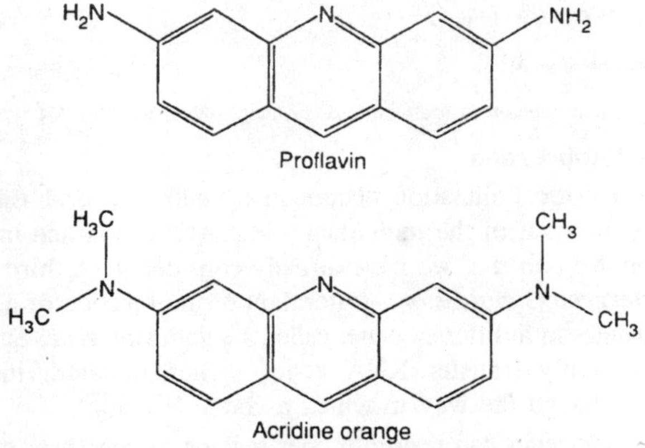

H_2N N NH_2

Proflavin

H_3C CH_3 N N N H_3C CH_3

Acridine orange

Fig. 3.17. Structure of two acridine dyes: proflavin and acridine orange.

Misalignment Mutagenesis

Additions and deletions in DNA can also come about by misalignment of a template strand and the newly formed (progeny) strand in a region containing a repeated sequence. For example, we expect the progeny strand to contain six adjacent adenines because the template strand contains six adjacent thymines. Misalignment of the progeny strand results in seven consecutive adenines: six thymines replicated, plus one already replicated but misaligned. Misalignment of the template strand results in five consecutive adenines because one thymine is not available in the template. Regions with long runs of a particular base may be very mutation prone. They may explain the "hot spots" observed by Benzer and others.

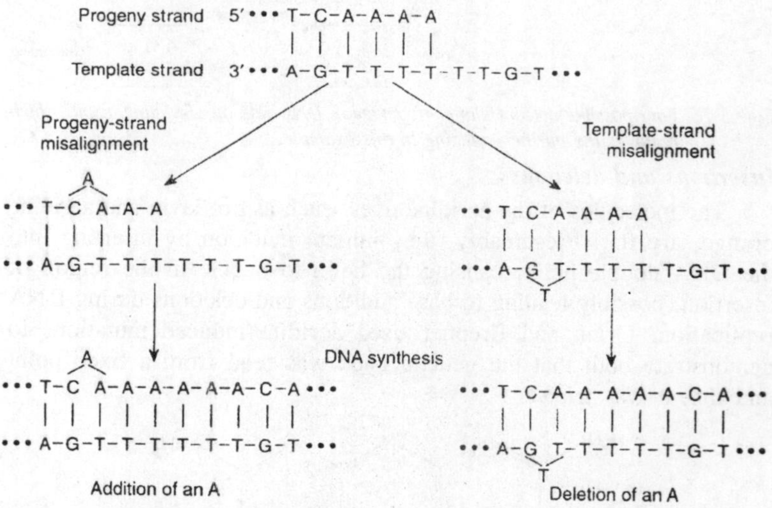

Fig. 3.18. Misalignment of a template or progeny strand during DNA synthesis.

Intergenic Suppression

When a critical mutation occurs in a codon, several routes can still lead to survival of the individual; simple reversion and intragenic suppression are two that we have already considered. A third route is through *intergenic suppression*—restoration of the function of a mutated gene by changes in a different gene, called a *suppressor gene*. Suppressor genes are usually transfer RNA genes. When mutated, intergenic suppressors change the way in which a codon is read.

Suppressor genes can restore proper reading to nonsense, missense, and frameshift mutations. *Nonsense mutations* convert a codon that originally specified an amino acid into one of the three nonsense codons. *Missense mutations* change a codon so that it specifies a different amino acid. Frameshift mutations, by additions or deletions of nucleotides, cause an alteration in the reading frame of codons. A

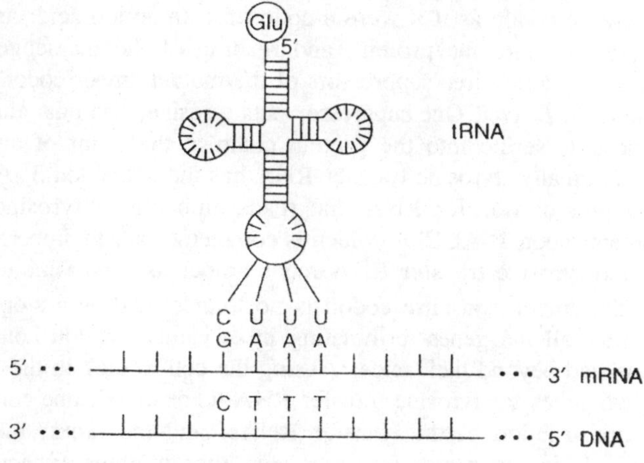

(a) Frameshift suppression

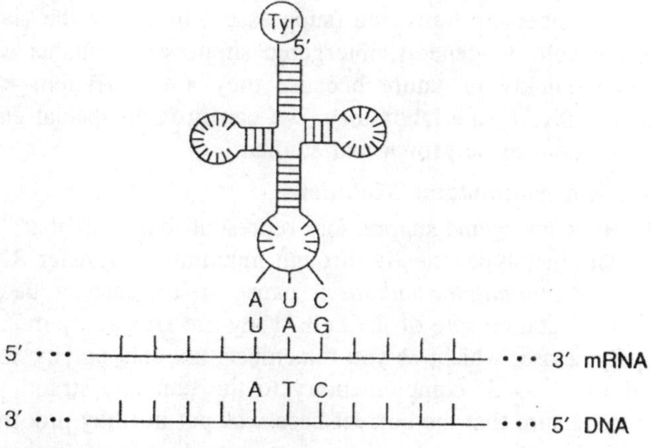

(b) Nonsense suppression

Fig. 3.19. Frameshift and nonsense suppression by mutant transfer RNAs. In (a) a thymine has been inserted into DNA, resulting in a frameshift. In (b), an amber mutation (UAG), which normally results in chain termination, is read as tyrosine by a mutant tyrosine transfer RNA that has the anticodon sequence complementary to the amber codon.

frameshift mutation, caused by the insertion of a single base, can be suppressed by a transfer RNA that has an added base in its anticodon. It reads four bases as a codon and thus restores the original reading frame.

The transfer RNA produced by a nonsense suppressor gene reads the nonsense codon as if it were a codon for an amino acid; an amino acid is placed into the protein, and reading of the messenger RNA continues. At least three suppressors of the mutant amber codon (UAG) are known in *E. coli*. One suppressor puts tyrosine, one puts glutamine, and one puts serine into the protein chain at the point of an amber codon. Normally, tyrosine transfer RNA has the anticodon 3'-AUG-5'. The suppressor transfer RNA that reads amber as a tyrosine codon has the anticodon 3'-AUC-5', which is complementary to amber. Hence, a mutated tyrosine transfer RNA reads amber as a tyrosine codon.

If the amber nonsense codon is no longer read as a stop signal, then won't all the genes terminating in the amber codon continue to be translated beyond their ends, causing the cell to die? In the tyrosine case, two genes for tyrosine transfer RNA were found; one contributes the major fraction of the transfer RNAs, and the other, the minor fraction. It is the minor-fraction gene that mutates to act as the suppressor. Thus, most messenger RNAs are translated normally, and most amber mutations result in premature termination, although a sufficient number are translated (suppressed) to ensure the viability of the mutant cell. In general, intergenic suppressor mutants would be eliminated quickly in nature because they are inefficient—the cells are not healthy. In the laboratory, we can provide special conditions that allow them to be grown and studied.

Mutator and Antimutator Mutations

Whereas intergenic suppressors represent mutations that "restore" the normal phenotype, mostly through mutation of transfer RNA loci, *mutator* and *antimutator mutations* cause an increase or decrease in the overall mutation rate of the cell. They are frequently mutations of DNA polymerase, which, as you remember, not only polymerizes DNA nucleotides 5' → 3' complementary to the template strand, but also checks to be sure that the correct base was put in (they proofread). If, in the proofreading process, the polymerase discovers an error, it can correct this error with its 3' → 5' exonuclease activity. Mutator and antimutator mutations sometimes involve changes in the polymerase's proofreading ability (exonuclease activity). Phage T4 has its own DNA polymerase with known mutator and antimutator mutants. Mutator mutants are very poor proofreaders (they have low exonuclease-to-polymerase ratios), and thus they introduce mutations throughout the phage genome. Antimutator mutants, however, have exceptionally efficient proofreading ability (high exonuclease-to-polymerase ratios) and, therefore, result in a very low mutation rate for the whole genome.

DNA Repair

Radiation, chemical mutagens, heat, enzymatic errors, and spontaneous decay constantly damage DNA. For example, it is estimated that several thousand DNA bases are lost each day in every mammalian cell due to spontaneous decay. Some types of DNA damage interfere with DNA replication and transcription. In the long evolutionary challenge to minimize mutation, cells have evolved numerous mechanisms to repair damaged or incorrectly replicated DNA. Many enzymes, acting alone or in concert with other enzymes, repair DNA. Repair systems are generally placed in four broad categories: damage reversal, excision repair, double-strand break repair, and postreplicative repair. Enzymes that process repair steps have been conserved during evolution. That is, enzymes found in *E. coli* have homologues in yeast, fruit flies, and human beings. However, eukaryotic systems are almost always more complex.

Damage Reversal

Ultraviolet (UV) light causes linkage, or *dimerization*, of adjacent pyrimidines in DNA. Although cytosine-cytosine and cytosine-thymine dimers are occasionally produced, the principal products of UV irradiation are thymine-thymine dimers. These can be repaired in several different ways. The simplest is to reverse the dimerization process and restore the original unlinked thymines.

In *E. coli*, an enzyme called DNA photolyase, the product of the *phr* gene (for *photoreactivation*), binds to dimerized thymines. When light shines on the cell, the enzyme breaks the dimer bonds with light energy. The enzyme then falls free of the DNA. This enzyme thus reverses the UV-induced dimerization. Another example of an enzyme that performs direct DNA repair is O^6-mGua DNA methyl-transferase, which removes the methyl groups from O^6-methylguanine, the major product of DNA-methylating agents. Although other repair mechanisms seem to be present in all organisms, photoreactivation is not; it is apparently absent in human beings.

Excision Repair

Excision repair refers to the general mechanism of DNA repair that works by removing the damaged portion of a DNA molecule.

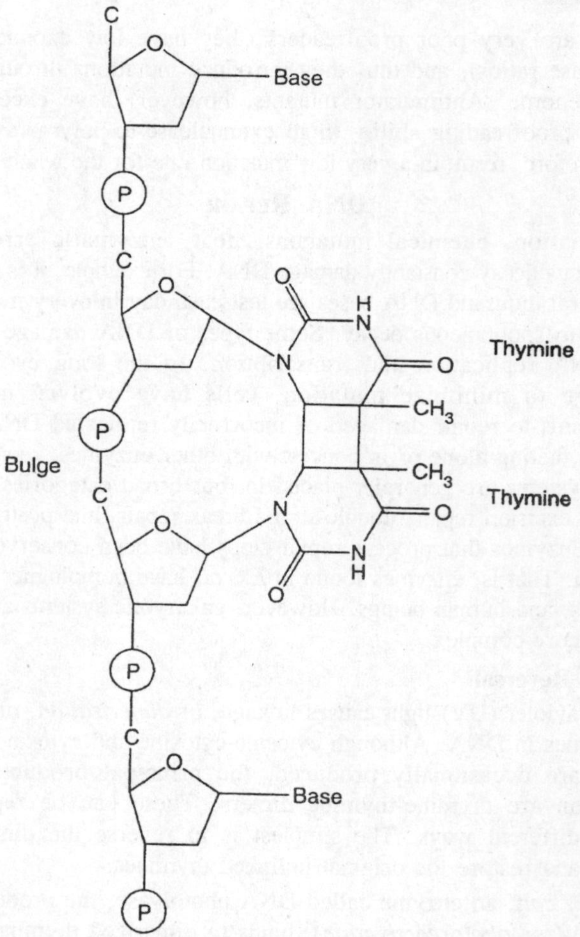

Fig. 3.20. UV-induced dimerization of adjacent thymines in DNA.

Various enzymes can sense damage or distortion in the DNA double helix. During excision repair, bases and nucleotides are removed from the damaged strand. The gap is then patched using complementarity with the remaining strand. We can broadly categorize these systems as base excision repair and nucleotide excision repair, which includes mismatch repair. We will discuss only the major repair pathways; others exist. Presumably, redundancy in repair has been selected for because of the critical need to keep DNA intact and relatively mutation free.

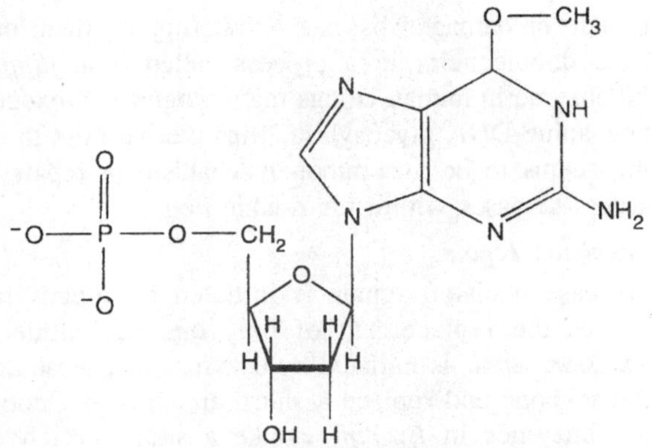

Fig. 3.21. The structure of O^5-methylguanine.

Base excision repair

A base can be removed from a nucleotide within DNA in several ways: by direct action of an agent such as radiation, by spontaneous hydrolysis, by an attack of oxygen free radicals, or by *DNA glycosylases*, enzymes that sense damaged bases and remove them. Currently, at least five DNA glycosylases are known. For example, uracil-DNA glycosylase, the product of the *ung* gene in *E. coli,* recognizes uracil within DNA and cleaves it out at the base-sugar (glycosidic) bond. The resulting site is called an AP (apurinic-apyrimidinic) site, because of the lack of a purine or pyrimidine at the site. An *AP endonuclease* then senses the minor distortion of the DNA double helix and initiates excision of the single AP nucleotide in a process known as *base excision repair*. The AP endonuclease nicks the DNA at the 5' side of the base-free AP site. A DNA polymerase then inserts a nucleotide at the AP site; an exonuclease, lyase, or phosphodiesterase enzyme then removes the base-free nucleotide. (Lyases are enzymes that can break C-C, C-O, and C-N bonds.) DNA ligase then closes the nick. The replacement of just one base occurs 80–90% of the time. In the remaining 10–20% of cases, several nucleotides may be removed, depending probably on which DNA polymerase (I or III) first repairs the site. In mammals, DNA polymerase β performs two roles in base excision repair: It both inserts a new base where the AP site was and also eliminates the AP nucleotide residue by exonuclease activity.

One question that concerned scientists was how the glycosylases gain access to the inappropriate or damaged bases within the double helix. Recently, it has been demonstrated that these enzymes remove the inappropriate or damaged bases by first flipping them out of the interior of the double helix in a process called *base flipping*. For example, the enzyme in human beings that recognizes 8-oxoguanine in DNA, 8-oxoguanine DNA glycosylase, flips the base out to excise it. Base flipping seems to be a common mechanism in repair enzymes that need access to bases within the double helix.

Nucleotide excision repair

Whereas base excision repair is initiated by glycosylases and usually involves the replacement of only one nucleotide residue, *nucleotide excision repair* is initiated by enzymes that sense distortions in the DNA backbone and replace a short stretch of nucleotides. For example, six enzymes in *E. coli* excise a short stretch of DNA containing thymine dimers if the dimerization is not reversed by photoreactivation. Two copies of the protein product of the *uvrA* gene (for ultraviolet light— UV—repair) combine with one copy of the product of the *uvrB* gene to form a $UvrA_2UvrB$ complex that moves along the DNA, looking for damage. (The complex has 5' to 3' helicase activity.) When the complex finds damage such as a thymine dimer, with moderate to large distortion of the DNA double helix, the $UvrA_2$ dimer dissociates, leaving the UvrB subunit alone. This causes the DNA to bend and attracts the protein product of the *uvrC* gene, UvrC. The UvrB subunit first nicks (hydrolyzes) the DNA four to five nucleotides on the 3' side of the lesion; next, the UvrC subunit nicks the DNA eight nucleotides on the 5' side of the lesion. (The three components, UvrA, UvrB, and UvrC, are together called the *ABC excinuclease*, for excision endonuclease.) The enzyme helicase II, the product of the *uvrD* gene, then removes the twelve-to thirteen-base oligonucleotide as well as UvrC. DNA polymerase I fills in the gap and, in the process, evicts the UvrB, and DNA ligase closes the remaining nick. This is another relatively simple system designed to detect helix distortions and repair them.

Like base excision repair, nucleotide excision repair is present in all organisms. In yeast, approximately twelve genes are involved, many in what is called the RAD3 group. In human beings, twenty-five proteins are involved; they remove twenty-seven to twenty-nine nucleotides, as compared to twelve to thirteen in *E. coli*.

Transcription and nucleotide excision repair are linked in eukaryotes. Transcription factor TFIIH is involved in repair of UV damage; it has helicase activity and is found in both processes. Since it has been shown that genes that are actively being transcribed are preferentially repaired, we can now envision a model in which transcription, when blocked by a DNA lesion like a thymine dimer, signals the formation of a repair complex, using TFIIH in both processes. In prokaryotes, RNA polymerase dissociates from the DNA in this circumstance, losing the nascent transcript. This would be inefficient in eukaryotes, whose genes are much longer and more expensive to transcribe; for example, the human dystrophin gene, defective in the disease Duchenne muscular dystrophy, is 2.4 million bases long and takes almost eight hours to transcribe. We believe that eukaryotic RNA polymerase II backs up when stalled at a DNA lesion and continues after the lesion is repaired, without losing the transcript. Much active research is going on in this area.

In human beings, the autosomal recessive trait *xeroderma pigmentosum* is caused by an inability to repair thymine dimerization induced by UV light. Persons with this trait freckle heavily when exposed to the UV rays of the sun, and they have a high incidence of skin cancer. There are seven complementation groups (loci *XPA-XPG*) whose protein products are involved in the first steps of nucleotide excision repair and whose defects cause xeroderma pigmentosum in human beings. One of them, *XPD,* is a component of TFIIH.

Excision repair triggered by mismatches is referred to as *mismatch repair*, which encompasses about 99% of all DNA repairs. As DNA polymerase replicates DNA, some errors are made that the proofreading polymerase does not correct. For example, a template G can be paired with a T rather than a C in the progeny strand. The GT base pair does not fit correctly in the DNA duplex. The mismatch repair system, which follows behind the replicating fork, recognizes this problem. This system, whose members in *E. coli* are specified by the *mutH, mutL, mutS,* and *mutU* genes, is responsible for the removal of the incorrect base by an excision repair process. (The genes are called *mut* for mutator because mutations of these genes cause high levels of spontaneous mutation in the cells. The *mutU* gene is also known as *uvrD.*) The mismatch repair enzymes initiate the removal of the incorrect base by nicking the DNA strand on one side of the mismatch.

You might wonder how the mismatch repair system recognizes the progeny, rather than the template, base as the wrong one. After all, in a mismatch, there are no defective bases—theoretically, either partner could be the "wrong" base. In *E. coli*, the answer lies in the methylation state of the DNA. DNA methylase, the product of the *dam* locus, methylates 5'-GATC-3' sequences, which are relatively common in the DNA of *E. coli*, at the adenine residue. Since the mismatch repair enzymes follow the replication fork of the DNA, they usually reach the site of mismatch before the methylase does. Template strands will be methylated, whereas progeny strands, being newly synthesized, will not be. Thus, the methylation state of the DNA cues the mismatch repair enzymes to eliminate the progeny-strand base for repair. After the methylase passes by, both strands of the DNA are methylated, and the methylation cue is gone.

The MutS protein, in the form of a homodimer—two copies of the same protein—finds the mismatch. MutL, also in the form of a homodimer, then binds, and together they find the methylation signal. They also activate the endonuclease MutH, which then nicks the unmethylated strand at the 3'-CTAG-5' recognition site, which can be one thousand to two thousand bases away from the mismatch. At the recognition site, the MutS-MutL tetramer loads the helicase MutU (UvrD), which then unwinds the nicked strand. Any one of at least four different exonucleases then attacks the unwound oligonucleotide. DNA polymerase III then repairs the gap, and DNA ligase seals it. This sequence of events highlights a common theme in DNA repair: Once a lesion is found, the damaged DNA has some protein bound to it until the repair is finished.

Our understanding of DNA damage and repair helps provide an answer to an evolutionary question—Why does DNA have thymine while RNA has uracil? If we live in an RNA world, in which RNA evolved first, why don't DNA and RNA both contain uracil? One answer is that a common damage to cytosine, spontaneous deamination, results in uracil. If uracil were a normal base in DNA, the conversion of cytosine to uracil by deamination would not leave any clue to a mismatch repair system that a mutation had occurred. Thus, thymine replaces uracil in DNA, since thymine is not confused with any other normal base in DNA by common spontaneous changes. In fact, cytosine, guanine, adenine, and thymine are not converted simply to any other of the bases in DNA. Hence, changes of these bases leave clues for the repair systems.

Double-Strand Break Repair

Some damage to DNA, such as that caused by ionizing radiation, is capable of breaking both strands of the double helix. When that happens, the cell uses one of two mechanisms to repair the broken ends: It can simply bring the ends back together (a process called *nonhomologous end joining*), or it can use a mechanism that relies on the nucleotide sequences of a homologous piece of DNA, such as a sister chromatid or a homologous chromosome. That method is called *homologydirected* recombination.

In nonhomologous end joining, a protein called Ku, a heterodimer of Ku70 and Ku80, binds to broken chromosomal ends. It then recruits a protein kinase (PK_{CS}); their interaction and the interaction with other proteins is stabilized by a scaffold protein called XRCC4 (for X-ray cross complementation group 4). The complex directs the annealing of the broken ends by DNA ligase IV. No particular sequence information is used, and if more than two broken ends are present, incorrect attachments can take place (e.g., translocations). The second method, homology-directed recombination, involves a second piece of DNA homologous to the broken piece. The method is very similar to our current model of DNA recombination and is discussed in the section entitled "Recombination" later in the chapter.

Postreplicative Repair

When DNA polymerase III encounters certain damage in *E. coli*, such as thymine dimers, it cannot proceed. Instead, the polymerase stops DNA synthesis and, leaving a gap, skips down the DNA to resume replication as far as eight hundred or more bases away. If allowed to remain, this gap will result in deficient and broken DNA. Since part of one strand is absent and the other has damage, there appears to be no viable template for replicating new DNA. However, the cell has two mechanisms to repair this gap: one uses polymerases that can replicate these lesions, and the other is a repair process that uses homologous DNA.

Originally, several proteins were known to facilitate the replication of DNA with lesions; they were believed to interact with the polymerase to make it capable of using damaged DNA as a template. We now know that these proteins are, in fact, polymerases that have the ability to replicate damaged DNA. In *E. coli*, polymerase V can copy damaged DNA. In yeast, polymerases η and ζ, also called REV3/7 and RAD30 polymerases, respectively, can also copy damaged DNA. Some of these polymerases are relatively error free; polymerase V and polymerases η put adenine-containing nucleotides opposite dimerized thymines. However, polymerases ζ and the *E. coli* polymerase IV, which also appears during times of damage, are error prone in their replicative roles. One possible reason for this is that the error-prone polymerases developed by evolutionary processes: They create mutations at a time when the cell might need variability. That is, DNA damage can occur when the environment is stressful for the cell; variability might help the cell survive. As we will see later, the cell can sense DNA damage and act appropriately.

In addition to using repair polymerases, the cell can use a second repair mechanism to replicate damaged DNA when the polymerase leaves a gap. A replication fork creates two DNA duplexes. Thus, an undamaged copy of the region with the lesion exists on the other daughter duplex. A group of enzymes, with one specified by the *recA* locus having central importance, repairs the gap. Since the repair takes place at a gap created by the failure of DNA replication, the process is called *postreplicative repair*. The *recA* locus was originally discovered and named in the recombination process. In fact, postreplicative repair is sometimes called recombinational repair, and it shares many enzymes with recombination.

RecA protein

The RecA protein has two major properties. First, it coats single-stranded DNA and causes that coated, single-stranded DNA to invade double-stranded DNA. By invasion, we mean that the single-stranded DNA attempts to form complementary base pairs with the antiparallel strand of the double-stranded DNA while displacing the other strand of that double helix. RecA continues to move the single-stranded DNA along the double-stranded DNA until a region of homology is found. The second major property of the RecA protein is that, when stimulated by the presence of single-stranded DNA, it causes autocatalysis of another repressor, called LexA, and thus initiates several sequences of reactions.

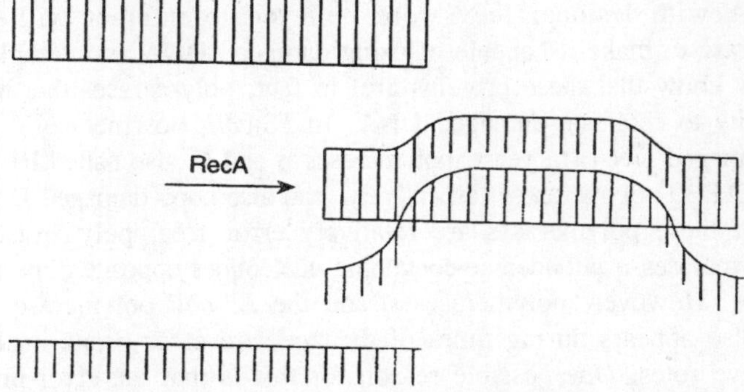

Fig. 3.22. One property of the RecA protein. It causes single-stranded DNA to invade double-stranded DNA and to move along the double-stranded DNA until a region of complementarity is found.

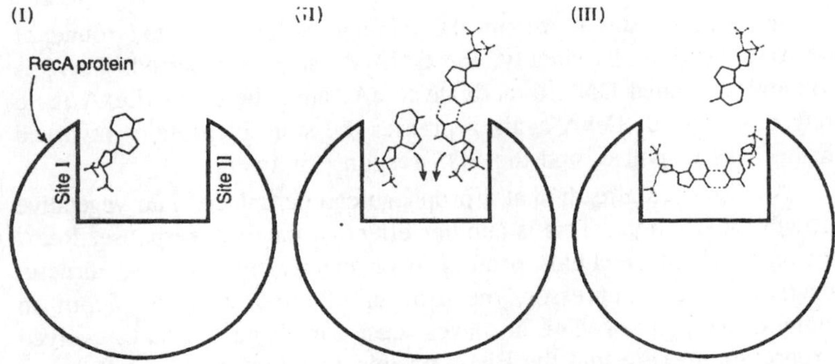

Fig. 3.23. A model of how the RecA protein can cause single-stranded DNA to invade a double-stranded molecule.

The RecA protein is responsible for filling a postreplicative gap in newly replicated DNA with a strand from the undamaged sister duplex. Gap-filling processes then complete both strands. The RecA protein is responsible for the damaged single strand invading the sister duplex. Endonuclease activity then frees the double helix containing the thymine dimer. DNA polymerase I and DNA ligase return both daughter helices to the intact state. The thymine dimer still exists, but now its duplex is intact, and another cell cycle is available for photoreactivation or excision repair to remove the dimer.

SOS response

Postreplicative repair is part of a cell reaction called the *SOS response*. When an *E. coli* cell is exposed to excessive quantities of UV light, other mutagens, or agents that damage DNA (such as alkylating or cross-linking agents), or when DNA replication is inhibited, gaps are created in the DNA. In the presence of this single-stranded DNA, the RecA protein interacts with the LexA protein, the product of the *lexA* gene. The LexA protein normally represses about eighteen genes, including itself. The other genes include *recA, uvrA, uvrB, and uvrD*; two genes that inhibit cell division, *sulA* and *sulB;* and several others. Each of these genes has a consensus sequence in its promoter called the *SOS box*: 5'-CTGX$_{10}$CAG-3' (where X$_{10}$ refers to any ten bases). The LexA protein normally binds at the SOS box, limiting the transcription of these genes. When single-stranded DNA activates RecA, RecA interacts with the LexA protein to trigger the autocatalytic properties of LexA. Transcription then follows from all the genes having an SOS box. The two inhibitors of cell division, the products of the *sulA* and *sulB* genes, presumably increase the amount of time the cell has to repair the damage before the next round of DNA replication. Eventually, the DNA damage is repaired. There is no single-stranded DNA to activate RecA, and, therefore, LexA is no longer destroyed. LexA again represses the suite of proteins involved in the SOS response, and the SOS response is over.

As discussed already that λ prophage can be induced into vegetative growth by UV light. This is another effect of the SOS response. RecA not only causes the LexA protein to be inactivated, but also directly inactivates the λ repressor, the product of the λ *cI* gene. From an evolutionary point of view, it makes sense for phage λ to have evolved a repressor protein that the RecA protein inactivates. As a prophage, λ is dependent on the survival of the host cell. When that survival might be in jeopardy, the prophage would be at an advantage if it could sense the danger and make copies of itself that could leave the host. One of these times might be when the host has suffered a lot of DNA damage. The SOS response is a signal to a prophage that the cell has received that damage. Hence, the prophage is induced when RecA acts as a protease; the λ repressor is destroyed, the cro protein becomes dominant, and vegetative growth follows. From an evolutionary perspective, the *E. coli* cell has not created an enzyme (RecA) that seeks out the λ repressor for the benefit of λ. Rather, the repressor has evolved for its own advantage to be sensitive to RecA.

Recombination

Although recombination, the nonparental arrangement of alleles in progeny, can come about both by independent assortment and crossing over, we are concerned here with recombination due to crossing over between homologous pieces of DNA (*homologous recombination*).

Recombination is a *breakage-and-reunion* process. Homologous parts of chromosomes come into apposition and are then reconnected in a crosswise fashion. This general model fits what we know about the concordance of recombination and repair: Both involve breakage of the DNA and a small amount of repair synthesis, and both involve some of the same enzymes.

Double-Strand Break Model of Recombination

In 1964, R. Holliday suggested a model of homologous recombination that involved simultaneous breaks in one strand each of the two double helices that were to cross over. In 1983, J. Szostak and colleagues put forth a different model, initiated by a double-strand break in one of the double helices. At first, this model was not considered seriously because a double-strand break was thought too dangerous a DNA lesion for cellular enzymes to create. However, we now know that the double-strand break model is generally correct, and we refer to the *Holliday junction* for an intermediate stage in the process. The model depends on DNA complementarity between the recombining molecules and is thus a model of great precision.

We begin with two double helices lined up as they would be, for example, in a meiotic tetrad, ready to undergo recombination. The first step of the process is a double-stranded break in one of the double helices. In eukaryotes, the protein Spo11 accomplishes this. The break is followed by 5' $\rightarrow$ 3' exonuclease activity to widen the gaps formed in the double helix and create 3' single-stranded tails. These tails are coated with RecA protein that then catalyzes the invasion of one of the single strands into the intact double helix in direct apposition. Repair of single-stranded DNA by DNA polymerase I and DNA ligase then replaces sections of previously digested DNA. At this point, there is no "lost" genetic material; however the two double helices are interlocked and need to be freed of each other. Before that happens, however, *branch migration* can take place, a process in which the crossover point can slide down the duplexes. In *E. coli*, the RuvAB complex, the product of the *ruvA* and *ruvB* genes that together form an ATP-dependent motor, moves the junction point. As the junction points move, they create heteroduplex DNA, places where the two strands of each double helix come from different original helices. These stretches have the potential to produce mismatches where the two chromatids differed originally. To resolve the cross-linked duplexes, a second cut at each junction is required.

Each of the two crossover points is a Holliday junction. If we open these junctions, we can see that each can be resolved in two different ways. (RuvC endonuclease, the protein product of the *ruvC* gene, resolves the Holliday junctions in *E. coli*. RuvC cuts the Holliday junction at the consensus sequence 5'[A or T]TT[G or C]-3'. The cut is on the 3' side of the two thymines.) Since there are two Holliday junctions per crossover, there are four potential combinations. Some of these combinations produce patches, where no recombination takes place among loci to the sides of the hybrid piece. Other combinations produce splices, where reciprocal recombination of loci takes place at the ends. The Holliday junctions can be seen in the electron microscope.

Bacterial Recombination

In bacterial recombination, a linear molecule recombines with a circular molecule. Usually, invading DNA originates in the linear molecule. The RecBCD protein, whose subunits are the products of the *recB, recC,* and *recD* loci, initiates the first steps in forming an invading linear DNA molecule. RecBCD is a helicase, an exonuclease, and an endonuclease.

The RecBCD protein enters a DNA double helix from one end and travels along it in an ATP-dependent process. As it travels along the DNA, it acts as a 3' → 5' exonuclease, degrading one strand of

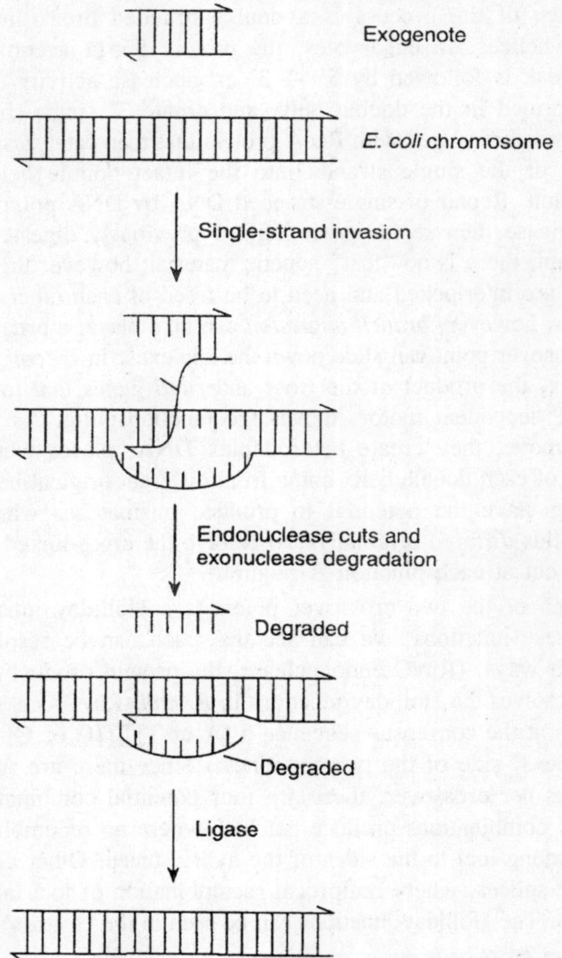

Fig. 3.24. Holiday model adapted to nonreciprocal incorporation of DNA into the E. coli chromosome.

the linear double helix. This process continues until RecBCD comes to a *chi site*, the sequence 5'-GCTGGTGG-3', which appears about a thousand times on the *E. coli* chromosome. RecBCD's recognition of that sequence attenuates its 3' → 5' exonuclease activity and enhances its 5' → 3' exonuclease activity, begun after an endonucleolytic cleavage. From that point on, RecBCD creates a 3' overhang or tail. That tail is coated by RecA and then invades the circular bacterial chromosome to initiate a crossover event. After this pairing, the unpaired segments of the double helix of the bacteria and the exogenote are both degraded. Finally, DNA ligase seals the circular double helix. The resulting hybrid DNA will then be open to mismatch repair that can restore either original base pairs or base pairs from the invading DNA.

Hybrid DNA

The result of bacterial recombination or meiotic recombination with branch migration is a length of hybrid DNA. This *hybrid DNA*, also called *heterozygous DNA* or *heteroduplex DNA*, has one of two fates, if we assume a difference in base sequences in the two strands. Either the heteroduplex can separate unchanged at the next cell division, or the cell's mismatch repair system can repair it. Without appropriate methylation cues, the mismatch repair system can convert the CA base pair to either a CG or a TA base pair. If TA were the original bacterial base pair, conversion to CG would be a successful recombination, whereas return of the CA to TA would be restoration rather than recombination.

Recombination in yeast, or any other eukaryote, generates two heteroduplexes. The repair process can cause *gene conversion*, the alteration of progeny ratios indicating that one allele was converted to another, a phenomenon seen in up to 10% of yeast asci. The mismatched AC will be changed to an AT or a GC base pair; the mismatched TG base pair will be changed to TA or CG. The result of the repair can be gene conversion in which an expected ratio of 2:2 ($a^- \; a^- \; a^+ \; a^+$) is converted to a 3:1 ratio ($a^- \; a^- \; a^- \; a^+$) or a 1:3 ratio ($a^- \; a^+ \; a^+ \; a^+$). If the heteroduplexes are not repaired, then a single cell generates both kinds of offspring after one round of DNA replication. Thus, the colony from the cell will be half wild-type (a^+) and half mutant (a^-). We see this phenomenon only in an Ascomycete fungus, such as yeast, in which all the products of a single meiosis remain together.

4

Gene Transcription and Modification

The chromosomal DNA of living organisms contains thousands of different genes. The information within those genesis accessed during gene expression. In this chapter, we will reconsider the first step in gene expression, transcription, in much detail. In particular, we will emphasize a molecular understanding of gene transcription and the modification of RNA molecules. For example, it is important to understand the sequences within genes that play a role in transcription. In this chapter we will examine these sequences. In addition, we will discuss the roles that different proteins play in the process of transcription. These proteins include RNA polymerase, transcription factors, and termination factors. Finally, we will consider the modifications that are necessary to produce functional RNA molecules.

Gene transcription is of the hottest topics in molecular biology. The transcription of genes is of central importance in a variety of research areas, including developmental biology, cancer biology, and biotechnology. In this chapter, we will focus on the basic components that are required to transcribe genes, in both prokaryotic and eukaryotic cells.

Overview of Transcription

The word transcription means the act or process of making a copy. In genetics, this term refers to the copying of a DNA sequence within a gene into an RNA sequence. DNA is a double-stranded molecule in which the two strands have complementary sequences according to the A—T/G—C rule. During transcription, one of the DNA strands is used as a template to make a complementary copy of

RNA. Except for the substitution of uradil in RNA for thymine in DNA, the base sequence in the RNA transcript is complementary to the *template strand* and identical to the DNA sequence in the opposite strand, known as the *coding strand*. In this introductory section, we will consider the general steps in the transcription process and the types of RNA transcripts that can be made.

Three Stages of Transcription are Initiation, Elongation, and Termination

Transcription occurs in three stages: *initiation*, *synthesis* of the RNA transcript (also called *elongation*), and *termination*. These steps involve protein—DNA interactions in which proteins such as RNA polymerase interact with DNA sequences. It is already stated that a gene contains two sequences that define the beginning and ending of transcription. These are known as the promoter and terminator, respectively. The initiation stage in the transcription process is a recognition step. The sequence of bases within the promoter region is recognized by proteins known as *transcription factors*. The specific binding of transcription factors to the promoter sequence identifies the starting site for transcription.

After the binding of transcription factors and RNA polymerase to the promoter region, the DNA strands must be separated for transcription to occur. One of the two strands can then be used as a template for the synthesis of a complementary strand of RNA. This synthesis occurs as RNA polymerase slides along the DNA, creating a small bubble-like structure known as the *open complex*. Eventually, the RNA polymerase reaches a termination sequence, which causes it and the newly made RNA transcript to dissociate from the DNA. This event is known as transcriptional termination.

RNA Transcripts have Different Functions

Once they are made, RNA transcripts play several different functional roles. A *structural gene* is transcribed, producing an RNA transcript known as messenger RNA (mRNA). The function of mRNA is to specify the amino acid sequence of a polypeptide. As will be described in the following chapter, the process of translation uses the codon sequence in mRNA to synthesize a polypeptide with a defined amino acid sequence. Well over 90% of all genes are structural genes.

In addition, several other genes encode RNAs that are never translated. The RNA transcripts from nonstructural genes have various important cellular functions. In some cases, the RNA transcript becomes part of a complex that contains both protein subunits and one or more

RNA molecules. Examples of protein—RNA complexes include ribosomes, signal recognition particles, spliceosomes, and certain enzymes such as *RNase P*. The function of spliceosomes and RNase P will be described later in this chapter.

Table 4.1. Functions of RNA molecules

Type of RNA	*Description*
mRNA	Messenger RNA (mRNA) encodes the sequence of amino acids within a polypeptide.
tRNA	Transfer RNA (tRNA) is necessary in the translation of mRNA.
rRNA	Ribosomal RNA (rRNA) is necessary in the translation of mRNA. rRNAs are components of ribosomes, which are composed of both rRNAs and protein subunits.
Other small RNAs	
7S RNA	7S RNA is necessary for targeting proteins to the endoplasmic reticulum. 7S RNA is a component of a complex known as signal recognition particle (SRP), which is composed of 7S RNA and six different protein subunits.
RNA of RNase P	RNase P is an enzyme that is necessary for processing bacterial tRNA molecules. The RNA is the catalytic component of this enzyme. RNase P is composed of a 350—410 nucleotide RNA and one protein subunit.
snRNA	Small nuclear RNA (snRNA) is necessary in the splicing of eukaryotic pre-mRNA. snRNAs are components of a spliceosome, which is composed of both snRNAs and protein subunits.
snoRNA	Small nucleolar RNA (snoRNA) is necessary in the processing of eukaryotic rRNA transcripts. snoRNAs are also associated with protein subunits. In eukaryotes, snoRNAs are found in the nucleolus, where rRNA processing and ribosome assembly occurs.

Transcription in Prokaryotes

Our molecular understanding of gene transcription initially came from studies involving bacteria and bacteriophages. Several early investigations focused on the production of viral RNA after bacteriophage infection. The first suggestion that RNA is derived from the transcription of DNA was made in the 1950s by Eliot Volkin and his colleagues at the Oak Ridge National Laboratory. They discovered that T2 bacteriophage infection of *Escherichia coli* resulted in the synthesis of RNA having a base composition similar to the DNA of the bacteriophage but different from the DNA of *E. coli*. These results were consistent with the idea that the bacteriophage DNA was being used as a template in the synthesis of bacteriophage RNA.

In 1960, Matthew Meselson at Cal Tech and Francois Jacob at the Pasteur Institute in Paris found that proteins are synthesized on ribosomes. One year later, Jacob and Jacques Monod proposed that a certain type of RNA acts as a genetic messenger (from the DNA to the ribosome) to provide the information for protein synthesis. They hypothesized that this RNA, which they called messenger RNA, is transcribed from the sequence within the DNA and then directs the synthesis of a particular polypeptide. In the early 1960s this was a remarkable proposal, considering that it was made even before the actual isolation and characterization of the mRNA molecules *in vitro*. The mRNA hypothesis was confirmed by Sydney Brenner at Cambridge in collaboration with Jacob and Meselson. They found that when a virus infects a bacterial cell, a virus-specific RNA is made that rapidly associates with preexisting ribosomes in the cell.

Since these pioneering studies of the 1950s and 1960s, we have learned a great deal about the molecular features of bacterial gene transcription. In this section, we will examine the three steps in the bacterial gene transcription process as they occur at the molecular level.

Prokaryotic Transcription is Initiated when RNA Polymerase Holoenzyme Binds at a Promoter Sequence

The type of DNA sequence known as the promoter gets its name from the idea that it "promotes" gene expression. More precisely; this sequence of nucleotides directs the precise location for the initiation of RNA transcription. A promoter is located just upstream from the site where transcription of a gene actually begins. When we describe the bases in a promoter sequence, by convention we number them relative to the transcription start site. In this system, the first base used as a template for RNA transcription is denoted +1. The bases preceding this are numbered in a negative direction. There is no base numbered 0. Therefore, the promoter region is labeled with negative numbers that describe the number of bases preceding the beginning of transcription.

Although the promoter may encompass a region that is several dozen nucleotides in length, there are short *sequence elements* within the promoter region that are particularly critical for promoter recognition. By comparing the sequence of DNA bases within many promoters, researchers have learned that certain sequences of bases are necessary to create a functional promoter. In many bacterial promoters, two sequence elements are important. These are located at approximately the —35 and —10 sites in the promoter region. The sequence at the —35 region is 5'—TTGACA—3', and the one at the —10 region is 5'—TATAAT—3'. The TATAAT sequence is sometimes called the Pribnow box after David Pribnow, who discovered it in 1975 while working at Harvard University.

The sequences at the —35 and —10 sites can vary among different genes illustrates the sequences found in several different prokaryotic promoters. The most commonly occurring bases within a sequence element form the *consensus sequence*. It is often times the sequence that is most efficiently recognized. For many bacterial genes, there is a good correlation between the rate of RNA transcription and the degree to which the —35 and —10 regions agree with their consensus sequences.

The enzyme that catalyzes the synthesis of RNA is *RNA polymerase*. It consists of multiple polypeptide subunits. In *E. coli*, the *core enzyme* is composed of four subunits; $\alpha_2\beta\beta'$. The association of a fifth subunit, known as *sigma factor*, with the core enzyme is referred to as the RNA polymerase *holoenzyme*.

The different subunits within the holoenzyme play distinct functional roles. The two α subunits are important in the proper assembly of the holoenzyme and in the process of binding to DNA. The β and β' subunits are critical in the catalytic synthesis of RNA. Overall, the core enzyme is necessary for RNA synthesis, and the primary role of the sigma factor is to recognize the promoter. Proteins, such as sigma factor, that influence the ability of RNA polymerase to transcribe a gene are known as *transcription factors*.

After RNA polymerase holoenzyme is assembled into its five subunits, it binds loosely to the DNA. It then scans along the DNA much as a train rolls down the tracks. When it encounters a promoter region, sigma factor recognizes the promoter elements at both the —35 and —10 positions. A region within the sigma protein that contains a *helix-turn-helix structure* is involved in a tighter binding to the DNA. Aipha-helices within the protein can fit into the major groove of the DNA double helix and form hydrogen bonds with the bases. Hydrogen bonding occurs between nucleotides in the —35 and —10 regions of the promoter and amino acid side chains in the helix-turn-helix portion of the sigma factor.

The process of transcription is initiated when the sigma factor within the holoenzyme has bound to the promoter region to form the *closed complex*. For transcription to begin, the double-stranded DNA must then be unwound into an open complex. This unwinding first occurs in the vicinity of the TATAAT box. The TATAAT box in the —10 region contains only A/T base pairs. A/T base pairs form only two hydrogen bonds, whereas G/C pairs form three. Therefore, it is easier to separate DNA in an A/T-rich region, since fewer hydrogen bonds must be broken. The formation of the open complex completes the initiation stage of transcription. The core enzyme may now slide down the DNA to synthesize a strand of RNA. During the synthesis stage, sigma factor is no longer necessary, and it dissociates from the holoenzyme.

RNA Transcript is Synthesized During the Elongation Stage

After the initiation stage of transcription is completed, the RNA transcript is actually made in the elongation stage of transcription. During the synthesis of the RNA transcript, RNA polymerase moves along the DNA, causing it to unwind. The DNA strand that is used as a template for RNA synthesis is called the template or noncoding strand. The opposite DNA strand is called the coding strand; it has the same sequence as the RNA transcript except that T in the DNA corresponds to U in the RNA. Within a given gene, only the template strand is used for RNA synthesis while the coding strand is never used. As it moves down the DNA, the open complex formed by the action of RNA polymerase is approximately 17 base pairs long. On average, the rate of RNA synthesis is about 43 nucleotides per second Behind the open complex, the DNA rewinds back into a double helix, thereby forcing the RNA behind the open complex to dissociate from the DNA.

The chemistry of transcription by RNA polymerase is similar to the synthesis of DNA via DNA polymerase. RNA polymerase always connects nucleotides in the 5' to 3' direction. During this process, RNA polymerase catalyzes the formation of a bond between the 5'— phosphate group on one nucleotide and the 3'—OH group on the previous nucleotide. The complemenarity rule is similar to the A—T/G—C rule, except that uracil substitutes for thymine in the RNA. In other words, RNA synthesis obeys an A_{RNA}–T_{DNA}/U_{RNA}–A_{DNA}/G_{RNA}–C_{DNA}/ C_{RNA}–G_{DNA} rule.

Transcription is Terminated by either an RNA-binding Protein or an Intrinsic Terminator

The end of RNA synthesis is referred to as termination. Prior to termination, the hydrogen bonding between the DNA and RNA within the open complex is of central importance in preventing dissociation of the RNA polymerase from the template strand. Termination occurs when this short RNA—DNA hybrid region is forced to separate, thereby releasing the newly made RNA transcript as well as the RNA polymerase. In *E. coil*, two different mechanisms for termination have been identified. For certain genes, a protein known as ρ (rho) is responsible for terminating transcription, a mechanism called ρ-dependent termination. For other genes, termination does not require ρ. This is referred to as ρ-independent termination. Both mechanisms wil; be described here.

In *ρ-dependent termination*, a sequence near the 3' end of the newly made RNA acts as a recognition site for the binding of the ρ protein. Rho protein functions as a helicase, an enzyme that can separate RNA—DNA hybrid regions. After the ρ recognition site is synthesized in the RNA, ρ binds to the RNA and moves in the direction of the RNA polymerase. A *termination sequence* near the 3' end of the gene causes RNA polymerase to pause in its synthesis of RNA. This allows ρ to catch up with RNA polymerase and break the hydrogen bonds between the DNA and RNA within the open complex. When this occurs, the completed RNA strand is separated from the DNA along with the RNA polymerase.

Rho-independent termination is facilitated by two sequence elements within the RNA. One element is a uracil-rich sequence located at the 3' end of the RNA. The second sequence is slightly upstream from the uradil-rich sequence, near the 3' end; it promotes the formation of a stem-loop (also called a hairpin) structure. A stem-loop structure can form due to complementary sequences within the RNA. These RNA stem-loops can form almost immediately after they are synthesized.

In the sequence of events, the formation of the stem-loop near the 3' end of the RNA causes the RNA polymerase to pause in its synthesis of RNA. At the time RNA polymerase pauses, the uracil-rich sequence in the RNA transcript is bound to the DNA template strand. As previously mentioned, the hydrogen bonding of the RNA to the DNA keeps the RNA polymerase clamped onto the DNA. However, the binding of this uracil-rich sequence to the DNA template strand is thought to be unstable, causing the RNA transcript to spontaneously fall off the DNA and terminate further transcription. Because this process does not require another protein to remove the RNA transcript from the DNA, the sequence elements that cause this type of termination are referred to as *intrinsic terminators*.

TRANSCRIPTION IN EUKARYOTES

Many of the basic features of gene transcription are very similar in prokaryotic and eukaryotic species. Much of our understanding of transcription has come from studies in *Saccharomyces cerevisiae* (baker's yeast) and higher eukaryotic species such as mammals. In general, gene transcription in eukaryotes is more complex than in their prokaryotic counterparts. Eukaryotic cells are larger and contain a variety of compartments known as organelles. This added level of cellular complexity dictates that eukaryotes contain many more genes encoding cellular proteins. In addition, higher eukaryotic species are multicellular, being composed of many different cell types. Multicellularity adds the requirement that genes be transcribed in the correct type of cell and during the proper stage of development. Therefore, in any given species, the transcription of the thousands of different genes that an organism possesses requires the appropriate timing and coordination. In this section, we will examine some features of gene transcription that are unique to eukaryotes.

Three Eukaryotic RNA Polymerases Transcribe Different Types of Genes

The genetic material within the nucleus of a eukaryotic cell is transcribed by three different RNA polymerase enzymes, designated I, II, and III. Each of the three RNA polymerases transcribes different categories of genes. *RNA polymerase I* transcribes all of the genes that encode ribosomal RNA (rRNA) except for the 5S rRNA. *RNA polymerase II* transcribes all of the structural genes and also certain snRNA genes. RNA polymerase II is responsible for the synthesis of all mRNA. *RNA polymerase III* transcribes all of the tRNA genes and the 5S rRNA gene.

All three RNA polymerases have many subunits. They contain two large sub units that are similar to the β and β' subunits of bacterial RNA polymerase. Eukaryotic RNA polymerases also contain eight or more additional subunits.

Transcription of Eukaryotic Genes is Initiated when RNA Polymerase and Transcription Factors Bind to a Promoter Sequence

In eukaryotes, the promoter sequence is more variable and often more complex than that found in prokaryotes. It is usually composed of a *core promoter* and *regulatory elements*. This chapter will focus primarily on the structure of the core promoter region. The core promoter is obligatory for transcription to take place. Similar to the function of a bacterial promoter, the core promoter provides the initial binding site for *general transcription factors* and RNA polymerase.

Common patterns of sequences found within the promoters of higher eukaryotes that are recognized by the three types of RNA polymerases. Short DNA sequences, which are usually called *elements* or *boxes*, are important for general transcription factors and RNA polymerase to assemble at the promoter region. For example, a common pattern of sequences found within the core promoter region that is recognized by RNA polymerase II. In the vicinity of the —25 position, a TATAAAA sequence called the *TATA box* is usually found. In addition, two other promoter elements are commonly found upstream from the TATA box. A sequence GGCCAATCT, called the *CAAT box*, and a sequence GGGCGG, called the *GC box*, may be located in the —50 to —100 region of structural genes in eukaryotes. However, the number and locations of GC boxes and CAAT boxes vary considerably among different eukaryotic structural genes.

The role of the TATA box differs from that of the CAAT and GC boxes. The TATA box is important in determining the precise starting point for transcription. If it is missing from the core promoter, the transcription starting point becomes undefined, and transcription may start at a variety of different locations. In contrast, the CAAT and GC boxes function as recognition sites that are bound by transcription factors and recruit RNA polymerase into the general vicinity of the promoter region.

Several general transcription factors are found in eukaryotes. These proteins interact in a fairly complicated way to recruit RNA polymerase to the core promoter region. Many general transcription factors have been identified, and their roles in transcriptional initiation have been studied. At present, the exact series of steps that occur during initiation is not entirely understood. Nevertheless, enough information is available to propose models that describe the events leading to the formation of the open complex. A model that describes the assembly of transcription factors and RNA polymerase II at the TATA box. As shown here, a series of interactions leads to the formation of the open complex. Transcription Factor IID (TFIID) first binds to the TATA box and thereby plays a critical role in the recognition of the Some large RNA transcripts are processed to smaller functional transcripts by enzymatic cleavage promoter. TFIID is composed of several subunits including TATA-binding protein (TBP), which directly binds to the TATA box, and several other proteins called TBP-associated factors (TAFs). After TFIID binds to the TATA box, it associates with TFIIA and TFIIB. TFIIB promotes the binding of RNA polymerase II. Transcription Factors IIE, IIF, and IIH are also bound to RNA polymerase, forming the closed complex.

To form the open complex, TFIIH hydrolyzes ATP and phosphorylates a do main in RNA polymerase II known as the carboxy terminal domain (CTD). These reactions release the contact between RNA polymerase II and TFIIB. TFIIH also functions as a helicase that breaks the hydrogen bonding between the double- stranded DNA and thereby promotes the formation of the open complex. After the open complex has formed, TFIIB, TFIIE, and TFIIH dissociate. RNA polymerase II is then free to proceed to the synthesis stage of transcription.

In vitro, when researchers mix together TBP, TFIIB, TFIIE, TFIIF, TFIIH, RNA polymerase, and a DNA sequence containing a TATA box, the DNA is transcribed into RNA. Therefore, these components are referred to as the *basal transcription apparatus*. In a living cell, however, additional components regulate transcription and allow it to proceed at a reasonable rate. TAFs (which are part of TFIID) and TFIIA are normally involved in the initiation of transcription. A multiprotein complex called *mediator* also binds to the CTD of RNA polymerase and regulates transcription. In addition, regulatory transcription factors bind to sequence elements in the vicinity of the promoter and influence the assembly of the basal transcription apparatus.

Nucleosome Structure is Disrupted so that the RNA Transcript can be Synthesized

The chemistry of the synthesis of RNA in eukaryotes is identical to that in prokaryotes. In eukaryotes, however, the DNA is packaged into a nucleosome structure. Since RNA polymerase is a very large enzyme compared with a nucleosome, the tight wrapping of DNA within a nucleosome is expected to inhibit the ability of RNA polymerase to transcribe the DNA. To circumvent this problem, the nucleosome structure is significantly perturbed during transcription.

Some studies suggest that the histones are completely displaced, whereas others suggest that they are loosened but remain attached to the DNA.

Different models have been proposed to explain how the histones might be partially or completely removed from the DNA during transcription. One possibility is that there are transcription factors that function in the removal of the histone proteins. Alternatively, it has been hypothesized that the movement of RNA polymerase along the DNA causes positive supercoiling ahead of the open complex and negative supercoiling behind it. Since the nucleosome structure is negatively supercoiled, positive supercoiling occurring in front of RNA polymerase would disrupt the nucleosome structure. Along these same lines, negative supercoiling occurring behind RNA polymerase would favor the reformation of a tight nucleosome structure.

RNA Modification

During the 1960s and 1970s, studies in bacteria established the physical structure of the gene. The analysis of bacterial genes showed that the sequence of DNA within the coding strand corresponds to the sequence of nucleotides in the mRNA. During translation, the sequence of codons in the mRNA is then read, providing the instructions for the correct amino acid sequence in a polypeptide. The one-to-one correspondence between the sequence of codons in the DNA coding strand and the amino acid sequence of the polypeptide has been termed the *colinearity* of gene expression. Based on these early results in bacteria, researchers concluded that modification of mRNA transcripts is unnecessary for gene expression.

The situation, however, rapidly changed in the late 1970s, when the tools be came available to study eukaryotic genes at the molecular level. The scientific community was astonished by the discovery that eukaryotic structural genes are not always colinear with their functional mRNAs. Instead, the coding sequences within many eukaryotic genes are separated by DNA sequences that are not translated into protein. The coding sequences are called exons, and the sequences that interrupt them are called intervening sequences or introns. During transcription, an RNA is made corresponding to the entire gene sequence. Subsequently, the sequences in the RNA that correspond to the gene introns are cut out, while the RNA sequences derived from the exons are connected, or spliced together. This process is called *RNA splicing*. Since the 1970s, it has become apparent that splicing is a common genetic phenomenon in eukaryotic species. Splicing occurs occasionally in prokaryotes as well.

Aside from splicing, research has also shown that RNA transcripts can be modified in several other ways. For example, rRNAs and tRNAs are synthesized as long transcripts that are cleaved into smaller functional pieces. In addition, most eukaryotic mRNAs have a cap attached to their 5' end and a tail attached at their 3' end. In this section, we will examine the molecular mechanisms that account for several types of RNA modifications.

Some Large RNA Transcripts are Processed to Smaller Functional Transcripts by Enzymatic Cleavage

In many cases, the RNA transcript initially made during gene transcription is processed into smaller functional RNA molecules. This involves cleavage of the RNA to make smaller pieces. The ribosomal RNA gene is transcribed by RNA polymerase I to make a long primary transcript, known as 45S rRNA. (The term 45S refers to the sedimentation characteristics of this transcript in Svedberg units.) Following the synthesis of the 45S rRNA, cleavage occurs at several points. Three of the fragments, termed 5.8S, 18S, and 28S rRNA, are functional rRNA molecules that bind to proteins within the ribosomes. In eukaryotes, the processing of 45S rRNA and the assembly of ribosomal subunits occur in a structure within the cell nucleus known as the nucleolus.

The production of tRNA molecules also requires processing. Like ribosomal RNA, tRNAs are synthesized as large precursor RNAs that must be cleaved at both the 5' and 3' ends to produce mature functional tRNAs (i.e., ones that bind to amino acids). This processing event has been studied extensively in *E. coli*. Interestingly, the cleavage occurs differently at the 5' end and the 3' end. At the 5' end, the pre cursor tRNA is recognized by an enzyme known as *RNase P*. This enzyme is an endonuclease that cuts the precursor tRNA. The action of RNase P produces the correct 5' end of the mature tRNA. At the 3' end, two different enzymes trim the tRNA precursor. First, an endonuclease cleaves the precursor RNA to remove a 170- nucleotide segment. Next, an exonuclease, designated *RNase D*, binds to the 3' end and digests the RNA in the 3' to 5' direction. When it reaches a CCA sequence, the exonuclease stops digesting the precursor RNA molecule. Therefore, all tRNAs in *E. coli* have a CCA sequence at their 3' ends. Finally, certain bases in tRNA molecules may be covalently modified to alter their structure.

As researchers studied tRNA processing, they found certain enzymatic features that were very unusual and exciting, changing the

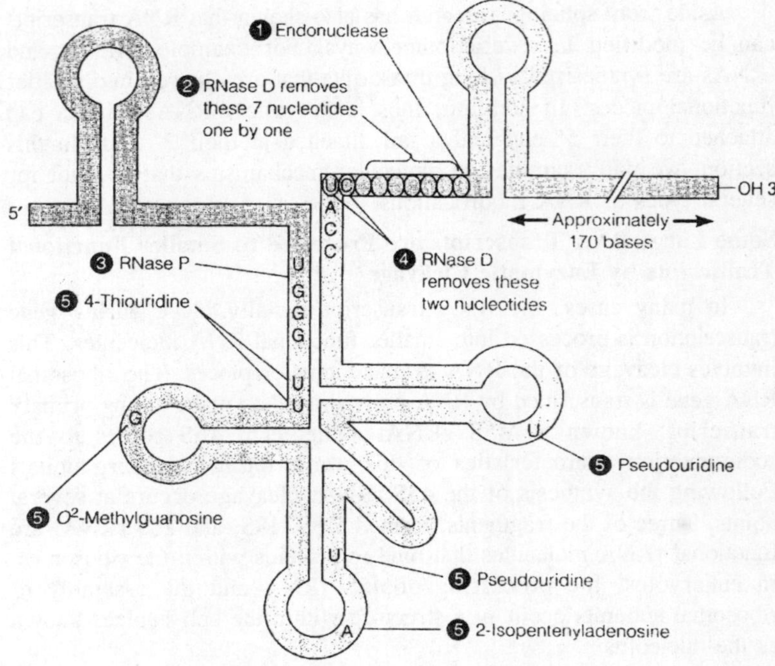

Fig. 4.1. The processing of tRNA molecules.

way biologists view the actions of enzymes. RNase P has been found to be an enzyme that contains both RNA and protein subunits. Surprisingly, Sidney Altman and colleagues at Yale University found that the RNA portion of this enzyme contains the catalytic ability to cleave the precursor tRNA. RNase P is an example of a *ribozyme*, which means that its catalytic ability is due to the action of RNA, not protein. Prior to the study of RNase P and the identification of self-splicing RNAs, biochemists had staunchly believed that only proteins could function as enzymes.

Leder and colleagues identified introns in the mouse β-globin gene in 1978

The discovery that tRNA and rRNA transcripts are processed to a smaller form did not seem unusual to geneticists and biochemists, because the enzymatic cleavage of RNA was similar to the cleavage that can occur for other macromolecules such as DNA and proteins. In sharp contrast, when splicing was detected in the 1970s, it was a novel concept. Splicing involves cleavage at two sites; an intron is removed; and—in a unique step—the remaining fragments are hooked hack together again.

Eukaryotic introns were first detected by comparing the base sequence of genes and mRNAs during viral infection of mammalian cells by adenovirus. This research was carried out by two groups: Philip Sharp and colleagues at MIT, and Richard Roberts and colleagues at Cold Spring Harbor. This pioneering observation led to the next question: "A introns a peculiar phenomenon that only occurs in viral genes, or are they found in chromosomal genes as well?"

In the late 1970s, several research groups, including those of Pierre Chambon in Strasbourg, France, Bert O'Malley at Baylor College of Medicine, and Phillip Leder at the National Institutes of Health (NIH), investigated the presence of introns in eukaryotic structural genes. The experiment described here is that of Leder, which used electron microscopy to identify introns in the β-globin gene.

In this experiment, Leder and his colleagues isolated the mRNA for the mouse β-globin gene. Beta-globin is a polypeptide that is a subunit of hemoglobin, the protein that carries oxygen in red blood cells. Prior to this work, the β-globin gene had been cloned. To detect introns within the cloned gene, Leder used a strategy involving *hybridization*. In this approach, the double-stranded DNA of the cloned β-globin gene was first denatured and mixed with mRNA for β-globin. Since this mRNA is complementary to the anticoding strand of the DNA, the anticoding strand and the mRNA will specifically bind to each other. This event is called hybridization. Later, when the DNA is allowed to renature, the binding of the mRNA to the anticoding strand of DNA prevents the two strands of DNA from forming a double helix. In the absence of any introns, the single-stranded DNA will form a loop. Since the RNA is displacing one of the DNA strands, this structure is known as an RNA displacement loop or *R loop*.

Leder and colleagues realized that a different type of R loop structure would be formed if the DNA contains an intron. For example, when mRNA hybridizes to a gene containing one intron, two single-stranded R loops will form that are separated by a double-stranded DNA region.

The intervening double-stranded region occurs because an intron has been spliced out of the mRNA, so that the mRNA cannot hybridize to this segment of the gene. As we will see, Leder and colleagues used electron microscopy to observe these kinds of structures in their experiment.

Hypothesis

The β-globin gene from the mouse contains introns.

Starting material: A cloned fragment of chromosomal DNA that contains mouse β-globin gene.

1. Isolate mature mRNA for the mouse β-glohin gene. Note: Globin mRNA is abundant in reticulocytes, which are immature red blood cells.
2. Mix the mRNA and cloned DNA together.
3. Separate the double-stranded DNA and allow the mRNA to hybridize. This is done using 70% formamide, 52°C, for 16 h.
4. Dilute the sample to decrease the formamide concentration. This allows the DNA to reform a double-stranded structure. Note: The DNA cannot form a double-stranded structure in regions where the mRNA has already hybridized.
5. Spread the sample onto a microscopy grid.
6. Stain with uranyl acetate and shadow with heavy metal.
7. View the sample under the electron microscope.

Interpreting the data

As seen in the electron micrograph, the mRNA hybridized to the DNA to form two R loops separated by a double-stranded DNA region. These data are consistent with the idea that the DNA of the β-globin gene contains an intron. Similar results were obtained by Chambon and O'Mailey for other structural genes. Since these early discoveries, introns have been found in many eukaryotic genes.

DNA sequencing methods developed in the late 1970s have permitted an easier and more precise way of detecting introns. Researchers can clone a fragment of chromosomal DNA that contains a particular gene. This is called a *genomic clone*. In addition, mRNA can be used as a starting material to make a copy of DNA known as *complementary DNA* or *cDNA*. The cDNA will not contain introns, because the introns have been previously removed during RNA splicing. In contrast, if a gene contains introns, a genomic clone for a eukaryotic gene will also contain introns. Therefore, a comparison of the DNA sequences of genomic and cDNA clones can provide direct evidence that a particular gene contains introns.

Three Different Splicing Mechanisms can Remove Introns

Since the 1970s, the investigations of many research groups have shown that most structural genes among higher eukaryotes contain one or more introns. Less commonly, introns can occasionally occur within tRNA and rRNA genes. At the molecular level, three different RNA splicing mechanisms have been identified. In all three cases, splicing leads to removal of the intron RNA and the covalent connection of the exon RNA by a phosphodiester linkage.

The chemistry of splicing among Group I and II introns is called *self-splicing*. It is given this name because the splicing event does not require the aid of other enzymes. Instead, the RNA functions as its own ribozyme. Group I introns that occur within the rRNA of *Tetrahymena* (a protozoan) have been studied extensively by Thomas Cech at the University of Colorado, Boulder. In this organism, the splicing process involves the binding of a single guanosine to a site within the intron This leads to the cleavage of RNA at the 3' end of exon 1. Next, the bond between a different guanine nucleotide (that is found within the intron sequence) and the 5' end of exon 2 is cleaved. This event allows the 3' end of exon 1 to then form a phosphodiester bond with the 5' end of exon 2. The intron RNA is subsequently degraded. In this example, the RNA molecule functions as its own ribozyme, since it splices itself without the aid of a catalytic protein.

In Group II introns, a similar splicing mechanism occurs, except the 2'—OH group on ribose found in an adenine nucleotide (already within the intron strand) begins the catalytic process Experimentally, Group I and II self-splicing can occur *in vitro* without the addition of any proteins. However, in a living cell (i.e., *in vivo*), proteins known as *maturases* often enhance the rate of splicing of Group I and II introns.

In eukaryotes, the transcription of structural genes produces a long transcript known as *pre-mRNA*, which is located within the nucleus. These large RNA transcripts are also known as *heterogenous nuclear RNA* or *hnRNA*. This pre-mRNA is usually altered by splicing and other modifications before it exits the nucleus. Unlike the Group I and II introns, which may undergo self-splicing, pre-mRNA splicing requires the aid of a multi component structure known as the *spliceosome*.

Table 4.2 describes the occurrence of introns among the genes of different species. The biological significance of Group I and II introns is not understood. By comparison, pre-mRNA splicing is a widespread phenomenon among higher eukaryotes. In mammals and higher plants, most structural genes have at least one intron that can be located anywhere within the gene. The functional significance of eukaryotic introns will be discussed later in this chapter.

Table 4.2. Occurrence of introns

Type of Intron	*Mechanism*	*Occurrence*
Pre-mRNA	Spliceosome	Very commonly found in structural genes with the nucleus of eukaryotes.
Group I	Self-splicing	Found in rRNA genes within the nucleus of Tetrahymena and other lower eukaryotes. Found in a few structural, tRNA, and rRNA genes within the mitochondrial DNA (fungi and plants) and chloroplast DNA. Found very rarely in tRNA genes within bacteria.
Group II	Self-splicing	Found in a few structural, tRNA, and rRNA genes within the mitochondrial DNA (fungi and plants) and in chloroplast DNA.

Most Eukaryotic Pre-mRNAs are Trimmed, Capped, Tailed, and Spliced

Before a pre-mRNA can be translated, several modifications must occur. A segment of RNA at the 5' or 3' end of the pre-mRNA may be removed by enzymatic cleavage. This process is known as *trimming*. Most mRNAs then have a 7-methylguanosine covalently attached to

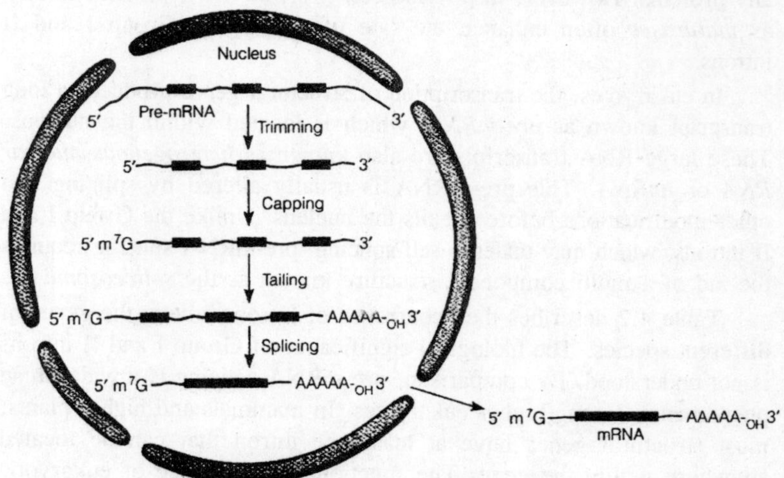

Fig. 4.2. Common modifications to eukaryotic pre-mRNA.

the 5' end. Such an attachment of this nucleotide to mRNA is known as *capping*; the cap is attached by a capping enzyme.

At the 3' end, most mRNAs have a string of adenine nucleotides, referred to as a polyA *tail*. The polyA tail appears to be important in the stability of mRNA and in the translation of polypeptides. However, it is not absolutely required for translation; some mRNAs without polyA tails are translated efficiently. To acquire a polyA tail, the pre-mRNA needs to contain a *polyadenylation sequence* near its 3' end. In higher eukaryotes, the consensus sequence is AAUAAA. This sequence is down stream (toward the 3' end) from the stop codon in the pre-mRNA. An endonuclease recognizes the polyadenylation sequence and cuts the pre-mRNA several nucleotides beyond the 3' end of the AAUAAA site. The fragment beyond the 3' cut is degraded. Next, an enzyme known as *polyA-polymerase* attaches many adenosine nucleotides. The length of the polyA tail varies among different mRNAs, from a few dozen to several hundred adenine nucleotides.

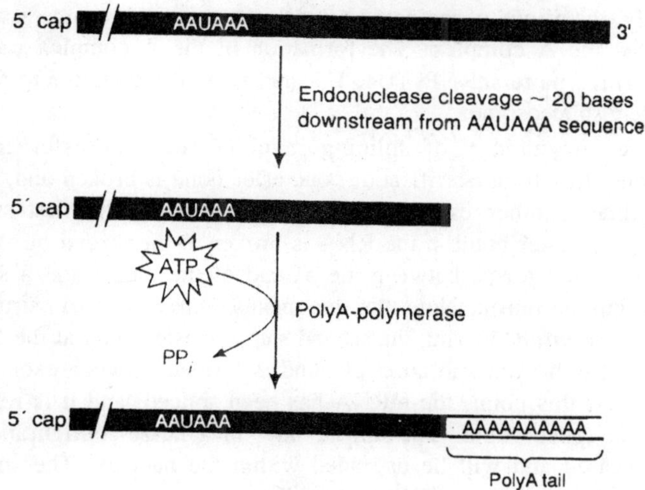

Fig. 4.3. Attachment of a polyA tail.

Besides trimming, capping, and tailing, a pre-mRNA may also be spliced be fore it is ready to exit the nucleus.

Pre-mRNA Splicing Occurs by the Action of a Spliceosome

As noted previously, the spliceosome is a large complex that splices pre-mRNA. It is composed of several subunits known as snRNPs (pronounced "snurps"). Each snRNP contains small nuclear RNA and a set of proteins. During splicing, the sub units of a spliceosome carry out several functions. First, spliceosome subunits bind to an intron sequence and precisely recognize the intron—exon boundaries. In addition, the spliceosome must hold the pre-mRNA in the correct configuration so that splicing can occur. And finally, the spliceosome catalyzes the chemical reactions that cause the introns to be removed and the exons to be covalently linked.

Intron RNA is recognized by the sequence of nucleotides within the intron and at the intron-exon boundaries. These sequences serve as recognition sites for the binding of the spliceosome. The most highly conserved bases are shown in bold.

First, the snRNP designated U1 binds to the 5' end of the intron at its boundary with the 3' end of the preceding exon, known as the *splice donor* site. The binding of U1 snRNP creates a *commitment complex* that is destined to proceed through the splicing cycle. This is followed by the binding of U2 to the branch site of the intron. Although not shown in this figure, U2 also recognizes the splice acceptor site (namely, the 5' end of the next exon); this step in spliceosome assembly is called the A complex. The formation of the A complex requires ATP. Three more snRNPs (U4, U5, and U6) then assemble to form a complete spliceosome.

The enzymology of splicing involves two transesterification reactions. In a transesterification, one ester bond is broken and, at the same time, another ester bond is formed. During the first step in splicing, an ester bond in the RNA is broken at the 3' end of exon 1. An ester bond forms between the 5' end of the intron and a second site within the intron. Note that this makes a circular intron structure, known as a *lariat*. During the second step, an ester bond at the 5' end of exon 2 is broken and an ester bond is formed between exon 1 and exon 2. At this point, the mRNA has been spliced, and it is released from the spliceosome. The intron, now in a lariat configuration, is also released and will be degraded within the nucleus. The snRNPs can then be recycled to splice more introns.

Alternative Splicing Allows Different Proteins to be Made from the Same Structural Gene

When it was first discovered, the phenomenon of splicing seemed like a rather wasteful process. During transcription, energy is used to synthesize intron sequences. Likewise, energy is also used to remove introns via a large spliceosome complex. This observation troubled many geneticists, because evolution tends to select against wasteful processes. Therefore, instead of simply viewing splicing as a wasteful process, many geneticists expected to find that pre-mRNA splicing has one or more important biological roles. In recent years, one very important biological advantage has become apparent. This is *alternative splicing*, which refers to the fact that a pre-mRNA can be spliced in more than one way.

To understand the biological effects of alternative splicing, remember that the sequence of amino acids within a polypeptide determines the structure and function of a protein. Alternative splicing produces two (or more) polypeptides with minor differences in their amino acid sequences, leading to subtle changes in their functions. In most cases, the alternative versions of the protein will have similar functions, because much of their amino acid sequences are identical to each other. Nevertheless, alternative splicing produces small differences in amino acid sequences that will provide each polypeptide with its own unique characteristics.

The biological advantage of alternative splicing is that two (or more) different polypeptide sequences can be derived from a single gene. This allows an organism to carry fewer genes in its genome. An example of alternative splicing for a gene that encodes a protein known as α-tropomyosin. This protein functions in the regulation of cell contraction. It is located along the thin filaments found in smooth muscle cells (e.g., in the uterus and small intestine) and striated muscle cells (e.g., in cardiac and skeletal muscle). Alpha-tropomyosin is synthesized in many types of nonmuscle cells as well. Within a multicellular organism, different types of cells must regulate their contractibility in subtly different ways. One way that this may be accomplished is to produce different forms of α-tropomyosin by alternative splicing.

The intron—exon structure of the rat α-tropomyosin gene and the alternative ways that the pre-mRNA can be spliced. The alternatively spliced versions of α-tropomyosin mRNA produce α-tropomyosin proteins that differ slightly from each other in their structure and function. All of the alternatively spliced versions of this mRNA contain exons 1, 4—6, and 9—10. Presumably, these exons encode polypeptide segments of the α-tropomyosin protein that are necessary for its general structure and function. However, the other exons are variable in different α-tropomyosin mRNAs. These alternative exons subtly change the function of α-tropomyosin to meet the needs of the cell type in which they are found. The specific pattern of splicing is usually uniform in any given cell. For example, the α-tropomyosin mRNA found in smooth muscle contains exon 2, which is not present in other cell types. Presumably exon 2 encodes a segment of the α-tropomyosin protein that slightly alters its function to make it suitable for smooth muscle cells. In another example, the splicing pattern of α-tropomyosin mRNA that contains exons 1, 3—6, 8—10, and 13 occurs in nonmuscle cells but not in smooth muscle, striated muscle, or brain cells.

Interestingly, certain abnormalities are associated with alternative splicing. For example, a *hepatoma* is a disease condition in which a liver cell becomes cancerous. A While normal liver cells do not transcribe the α-tropomyosin gene at a significant rate, a laboratory hepatoma cell line has been shown to produce an alternatively spliced version of α-tropomyosin that contains exon 7.

Nucleotide Sequence of RNA can be Modified by RNA Editing

The term *RNA editing* refers to a change in the nucleotide sequence of an RNA molecule that involves additions or deletions of particular bases, or a conversion of one type of base to another (e.g., a cytosine to a uracil). In the case of mRNAs, editing can have various effects, such as generating start codons, stop codons, and additional coding sequences.

The phenomenon of RNA editing was first identified in the mid-1980s in trypanosomes, the protozoa that cause sleeping sickness. As with the discovery of RNA splicing, the initial finding of RNA editing was met with great skepticism. Since that time, however, RNA editing has been shown to occur in various organ isms and in a variety of ways, although its functional significance remains unclear. Over the next decade or so, it will be interesting to see if RNA editing is a wide spread occurrence or if it only occurs in a few selected cases.

The molecular mechanisms of RNA editing have been the subject of many investigations. In the specific case of trypanosomes, the editing process involves the nucleotides of a *guide RNA* that directs the addition of one or more uracil nucleotides into the mRNA. The guide mRNA has two important characteristics: (1) its 5' anchor is complementary to the mRNA that is to be edited; and (2) it has uracil nucleotides at its 3' end. During the editing process, the 5' anchor binds to the mRNA. The mRNA is cleaved at a defined location, and the 3' end of the guide RNA becomes displaced from the mRNA. An enzyme called *terminal U-transferase* then attaches uracil-containing nucleotides. Finally, *RNA ligase* rejoins the two pieces of mRNA, resulting in an edited mRNA with additional uradil-containing nucleotides.

A different mechanism occurs for mammalian RNA editing that involves changes of one type of base to another. For example, an mRNA that encodes a protein called *apolipoprotein B* is edited so that a single C is changed to a U. This converts a glutamine codon (CAA) to a stop codon (UAA) and thereby results in a shorter apolipoprotein, termed B48. In this case, therefore, RNA editing produces an apolipoprotein B with an altered structure. Therefore, RNA editing can produce two proteins from the same gene, much like the phenomenon of alternative splicing described earlier in this chapter.

The apolipoprotein B mRNA editing process occurs within the nucleus of the cell at the same time that the pre-mRNA is being spliced. The C to U conversion is catalyzed by a complex known as an *editosome*, which recognizes a small region adjacent to the editing site and then modifies the cytosine base. The term editosome has also been used to describe the protein machinery that may be necessary for editing via guide RNA.

Concluding Remarks

Transcription is the process of A synthesis. Different types of A transcripts can be made. These include mRNA that specifies the sequence of amino acids within a polypeptide, and tRNA and rRNA, which are necessary to translate mRNA. In addition, several other small RNA molecules, such as snRNA, snoRNA, 7S RNA, and the RNA component of RNase P, have various functions in the cell.

The synthesis of RNA during gene transcription is a three-step process: *initiation*, *elongation*, and *termination*. During the initiation stage, specific sequences within the promoter region are recognized by proteins that bind to DNA. In the case of many bacterial promoters, the —35 and —to sequences are recognized by the *sigma factor* that is part of the *RNA polymerase* holoenzyme. To begin the *synthesis* of a transcript, the DNA must be converted to an *open complex* by denaturing a small double-stranded region. At this point, the RNA polymerase can slide down the DNA, synthesizing a complementary RNA molecule. At the end of the gene, a termination signal is found. In bacteria, there are two different ways that termination signals can promote transcriptional termination. In *ρ-dependent termination*, a protein known as ρ is responsible for dissociating the RNA transcript and RNA polymerase. In *ρ-independent* or *intrinsic termination*, the formation of a stem-loop within the RNA causes RNA polymerase to stall and eventually fall off the DNA.

Transcription in eukaryotes is similar to, but more complex, than that in prokaryotes. In eukaryotes, there are three RNA polymerases, designated I, II, and III, that transcribe different types of genes. The initiation of transcription of *structural genes* requires several *transcription factors* that recognize sequence elements such as *CAAT* and *GC boxes* and assemble at a *TATA box* in the —25 region. Transcription is initiated by a complicated assembly process. During elongation, the histone proteins must be completely or partially removed so that RNA polymerase can transcribe the template DNA.

Following transcription, a newly made RNA transcript may be modified in various ways. For example, rRNA and tRNA transcripts are processed to smaller forms by endo- and exonucleases. Some genes contain introns that must be removed from the RNA via *splicing*. Group I and Group II introns are self-splicing, which means that they catalyze their own removal. These types of introns are found in the rRNA genes of Tetrahymena and other simpler eukaryotes, in organelle DNA, and occasionally in bacterial genes. In contrast, introns are commonly found in eukaryotic pre-mRNA, where they are removed by a multimeric complex known as a spliceosome. Besides splicing, eukaryotic pre-mRNA is also trimmed, capped at its 5' end, and a polyA tail may be attached to its 3' end. It is particularly interesting that many eukaryotic pre mRNAs can be *alternatively spliced*. The net result of alternative splicing is that a single gene can produce more than one type of polypeptide. Similarly, the sequence of bases within RNA may be modified by *RNA editing*.

Experimentally there are several methods for studying the transcription of genes at the molecular level. For example, the quantity of an RNA transcript can be determined by Northern blotting, in which RNA transcripts are identified through a DNA—RNA hybridization technique. This method can also be used to detect alternative splicing of mRNAs.

Our understanding of the sequences required for transcription, such as promoter sequences, has come from the isolation of mutants in which the promoter sequence was altered. More recently, the technique of site-directed mutagenesis has been developed to alter gene sequences. Researchers now can examine how an alteration in a gene sequence affects genetic processes such as transcription.

The binding and assembly of proteins during transcription can be examined in a variety of ways. Similarly, researchers have succeeded in developing in vitro systems to study transcription. In this way, they have identified the basal transcriptional apparatus needed for transcription to occur.

In this chapter, we have also considered how hybridization and electron microscopy can be used to detect introns. In this approach, the hybridization of mRNA to a DNA sequence blocks the DNA ability to re-form double-stranded regions. Using an electron microscope to examine the pattern of single-stranded DNA and double-stranded DNA regions, the hybridization of RNA to DNA allows us to determine if a gene has introns. As mentioned at the end, DNA sequencing methods allow an easier and more precise way of detecting introns. A comparison of the DNA sequences of genomic and cDNA clones can determine if a particular gene contains introns.

5

TRANSLATION OF mRNA

The synthesis of cellular proteins occurs via the translation of the codons with in mRNA into a sequence of amino acids within a polypeptide. This chapter reexamines translation with an eye toward the molecular level. During the past few decades, the concerted efforts of geneticists, cell biologists, and biochemists have profoundly advanced our understanding of translation. Even so, many questions remain unanswered, and this topic continues to be an exciting area of research investigation.

In this chapter, we will discuss the current state of knowledge regarding the molecular features of mRNA translation. A variety of cellular components play important roles in translation. These include many different proteins, RNAs, and small molecules. In the first section of this chapter, we will begin by examining the structure and function of tRNA molecules, which act as the translators of the genetic information within mRNA. Later, we will consider the composition of ribosomes, and then examine the stages of translation in prokaryotic and eukaryotic cells.

tRNA STRUCTURE AND FUNCTION

Biochemical studies of protein synthesis and tRNA molecules began in the 1950s. As work progressed on protein translation, it became evident that RNA molecules are involved in the incorporation of amino acids into growing polypeptides. Francis Crick and Mahion Hoagland were the first to propose that the position of an amino acid within a polypeptide chain is determined by the bonding between the mRNA and a tRNA carrying a specific amino acid. This idea, known as the *adaptor hypothesis*, suggested that the tRNA plays a direct role in the recognition of the codons within the mRNA. In particular, this hypothesis proposed that a tRNA has two functions: recognizing a three-base codons sequence in mRNA, and carrying an amino acid that is specific for that codon. In this section, we will begin by examining the general function of tRNA molecules and describe an experiment that was critical in supporting the adaptor hypothesis. We will then explore some of the important structural features that underlie tRNA function.

The Function of a tRNA Depends on the Specificity between the Amino Acid it Carries and its Anticodon

T-RNA molecules recognize the codons within an mRNA and carry the correct amino acids to the site of polypeptide synthesis. During mRNA—tRNA recognition, the anticodon in a tRNA molecule binds to a codon in mRNA due to their complementary sequences. Importantly, the anticodon in the tRNA corresponds to the amino acid that it carries. For example, if the anti codon in the tRNA is 3'—AAG—5', it is complementary to a 5'—UUC—3' codon. According to the genetic code, the UUC codon specifies phenylalanine. Therefore, the tRNA with a 3'-AAG-5' anticodon must carry a phenylalanine. As another example, if the tRNA has a 3'—GGC—5' anticodon, it is complementary to a 5'—CCG—3' codon, which specifies proline. This tRNA must carry proline.

The genetic code has 3 stop codons and 61 different codons that specify the 20 amino acids. Therefore, to synthesize proteins, a cell must produce many different tRNA molecules that have specific anticodon sequences. To do so, the chromosomal DNA contains many distinct tRNA genes that encode tRNA molecules with different

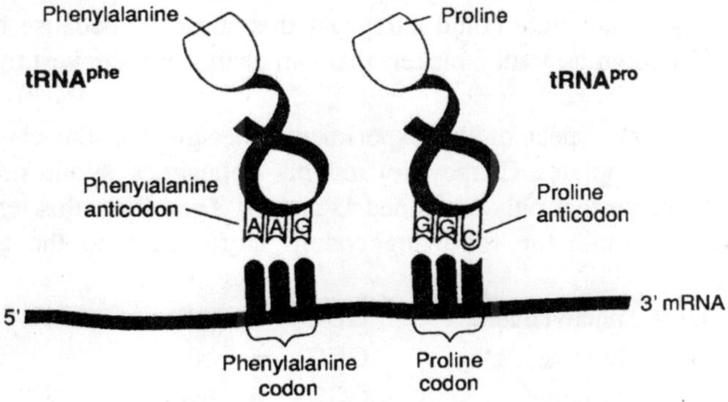

Fig. 5.1. Recognition between tRNAs and mRNA.

sequences. According to the adaptor hypothesis, the anticodon in a tRNA always specifies the type of amino acid that it carries. Due to this specificity, tRNA molecules are named according to the type of amino acid that they bear. For example, a tRNA that carries an phenylalanine is described as $tRNA^{phe}$, while a tRNA that carries proline is $tRNA^{pro}$.

Chapeville and Colleagues Tested the Adaptor Hypothesis of tRNA Function

In 1962, Francois Chapeville, in collaboration with several colleagues at the Rockefeller Institute, Johns Hopkins University, and Purdue University, con ducted an experiment that was aimed at testing the adaptor hypothesis. The technical strategy was similar to that of the Nirenberg and Ochoa experiments, which deciphered the genetic code. In this approach, a translation system was used in which polypeptides were synthesized *in vitro* by isolating cell extracts that contained the components necessary for translation. These components include ribosomes, tRNAs, and other translation factors. An *in vitro* translation system can be used to investigate the role of specific factors by adding a particular mRNA template and varying individual components required for translation.

According to the adaptor hypothesis, the amino acid attached to a tRNA is not directly involved in codon recognition. Chapeville reasoned that if this were true, the alteration of an amino acid already attached to a tRNA should cause that altered amino acid to be incorporated into the polypeptide instead of the normal amino acid. For example, consider a $tRNA^{cys}$ that carries the amino acid cysteine. If the attached cysteine is changed to an alanine, then this $tRNA^{cys}$ will insert an alanine into a polypeptide where it would normally put a cysteine. Fortunately, Chapeville could carry out this strategy, because he had a reagent, known as Raney nickel, that can chemically convert cysteine to alanine.

An elegant aspect of the experimental design was the choice of the mRNA template. Chapeville and his colleagues synthesized an mRNA template that only contained U and G. Therefore, this template could only contain the following codons (refer back to the genetic code):

UUU = phenylalanine   GUU = valine
UUG = leucine   GUG = valine
UGU = cysteine   GGU = glycine
UGG = tryptophan   GGG = glycine

Among the eight possible codons, there is one cysteine codon, but there are no ala nine codons that can be formed from a polyUG template.

Hypothesis

Codon recognition is dictated only by the tRNA; the chemical structure of the amino acid attached to the tRNA does not play a role.

Testing the hypothesis

Starting material: A bacterial cell extract containing the components required for translation, such as ribosomes, tRNAs, etc.

1. Isolate an extract containing tRNAs. Note: This drawing only emphasizes $tRNA^{cys}$ even though the extract contains all types of tRNAs. When the extract is isolated, a substantial proportion of the tRNAs do not have an attached amino acid. The extract also contains enzymes that attach amino acids to tRNAs.
2. Add amino acids, including radiolabeled cysteine, to the extract. An enzyme within the extract will specifically attach the radiolabeled cysteine to $tRNA^{cys}$. The other tRNAs will have unlabeled amino acids attached to them.
3. In one tube, treat the tRNAs with Raney nickel. This removes the —SH group from cysteine, converting it to alanine. In the control tube, do not add Raney nickel.
4. Add the tRNAs to the other components that are necessary for translation: ribosomes, translation factors, and so forth.
5. Add polyUG mRNA as a template. As noted, polyUG contains one cysteine codon but no alanine codons.
6. Allow translation to proceed.
7. Isolate the newly made polypeptides by precipitating them with trichloroacetic acid and then isolating the precipitated polypeptides on a fiter.
8. Hydrolyze the polypeptides to their individual amino acids by treatment with a solution containing concentrated hydrochloric acid.
9. Run the sample over a column that separates cysteine and alanine. Separate into fractions. Note: Cysteine runs through the column more quickly and comes out in fraction 3. Ahnine comes out later, in fraction 7.
10. Determine the amount of radioactivity in the fractions that contain alanine and cysteine.

Data

Conditions	*Cysteine*	*Alanine*	*Total*
Control, untreated tRNA	2835	83	2918
Raney nickel-treated tRNA	990	2020	3010

Interpreting the data

In the control sample, nearly all the radioactivity was found in the fraction containing cysteine. This was expected, since the only radiolabeled amino acid added to the tRNAs was cysteine. The low radioactivity (83 counts per minute) in the alanine fraction probably represents contamination of this fraction by a small amount of cysteine. By comparison, when the tRNAs were treated with Raney nickel, a substantial amount of radiolabeled alanine became incorporated into polypeptides. This occurred even though the mRNA template did not contain any alanine codons.

These results are consistent with the explanation that a $tRNA^{cys}$ carrying ala- nine instead of cysteine incorporated alanine into the synthesized polypeptide. Thus, these observations indicate that the codons in mRNA are identified directly by the tRNA, with the attached amino acid playing no role in codon recognition.

Note that the Raney nickel—treated sample still had 990 cpm of cysteine incorporated into polypeptides. This is about one-third of the total amount of radioactivity (namely, 990/3010). In other experiments conducted in this study, the researchers showed that the Raney nickel did not react with about one-third of the $tRNA^{cys}$. Therefore, this proportion of the Raney nickel—treated $tRNA^{cys}$ would still carry cysteine. This observation was consistent with the data shown here. Overall, the results of this experiment supported the adaptor hypothesis, indicating that tRNAs act as adaptors to carry the correct amino acid to the ribosome based on their anticodon sequence.

tRNAs Share Common Structural Features

To understand how tRNAs act as carriers of the correct amino acids during translation, researchers have examined the structural characteristics of these molecules in great detail. Though a cell makes many different tRNAs, all tRNAs share some common structural features. As originally proposed by Robert Hoiley at Cornell, the secondary structure of tRNAs exhibits a cloverleaf pattern. There are three stem-loop structures, a variable region, an acceptor stem, and a 3' single-stranded region. A conventional numbering system for the nucleotides within a tRNA molecule begins at the 5' end and proceeds toward the 3' end. Among different types of tRNA molecules, there are three variable regions that can differ in the number of nucleotides they contain. The anticodon is located in the second stem-loop region.

The actual three-dimensional or tertiary structure of tRNA molecules involves additional folding of the secondary structure. In the tertiary structure of tRNA, the stem-loop regions are folded into a much more compact molecule.

Interestingly, in addition to the normal A, U, G, and C nucleotides, tRNA molecules commonly contain modified nucleotides within their primary structures. Among many different species, researchers have found more than 60 different nucleotide modifications that can occur in tRNA molecules.

Aminoacyl-tRNA Synthetases Charge tRNAs by Attaching an Appropriate Amino Acid

To function correctly, each type of tRNA must have the appropriate amino acid attached to it. The enzymes that catalyze the attachment of amino acids to tRNA molecules are known as *aminoacyl-tRNA synthetases*. In most species, there are 20 different aminoacyl-tRNA synthetase enzymes, one for each of the 20 distinct amino acids. Each aminoacyl-tRNA synthetase is named for the specific amino acid it attaches to tRNA. For example, alanyl-tRNA synthetase recognizes a tRNA with an alanine anticodon (i.e., $tRNA^{ala}$) and attaches an alanine to it.

Aminoacyl-tRNA synthetases actually catalyze a two-step chemical reaction involving three different molecules: an amino acid, a tRNA molecule, and ATP. The first step of the reaction involves the activation of an amino acid by the covalent attachment of an AMP molecule. Pyrophosphate is re leased. During the second step, the activated amino acid is attached to the ribose located at the 3' end of the tRNA molecule in the acceptor stem; AMP is released. At this stage, the tRNA is called a *charged tRNA*. In a charged tRNA molecule, the amino acid is attached to the 3' end of the tRNA by an ester bond.

The ability of the amino acyl-tRNA synthetases to recognize tRNAs has some times been called the second genetic code. This recognition process is necessary to maintain the fidelity of genetic information. The frequency of error for aminoacyl-tRNA synthetases is less than 10^{-5}. In other words, the wrong amino acid will be attached to a tRNA less than once in 100,000 times! As you might expect, the anticodon region of the tRNA is usually important for recognition by the correct aminoacyl-tRNA synthetase. In studies of *Escherichia coli*

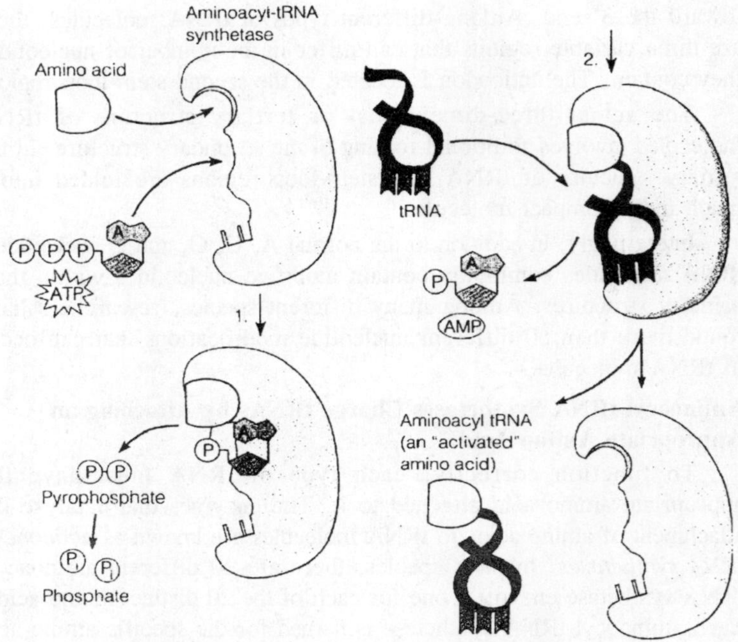

Fig. 5.2. Catalytic function of aminoacyl-tRNA synthetase.

synthetases, 17 of the 20 types of aminoacyl-tRNA synthetases recognize the anticodon region of the tRNA. However, other regions of the tRNA also appear to be important recognition sites. These include the acceptor stem and bases in the stem-loop regions.

As mentioned earlier in this chapter, tRNA molecules frequently contain bases within their structure that have been chemically modified. These modified bases can have important effects on tRNA function. For example, modified bases within tRNA molecules affect the rate of translation and the recognition of tRNAs by aminoacyl-tRNA synthetases. Positions 34 and 37 contain the largest variety of modified nucleotides; position 34 is the first base in the anticodon that matches the third base in the codon of mRNA. As discussed next, a modified base at position 34 can have important effects on codon—anticodon recognition.

Mismatches that Follow the Wobble Rule can Occur at the Third Position in Codon—Anticodon Pairing

After considering the structure and function of tRNA molecules, it is interesting to reexamine some of the subtle features of the genetic

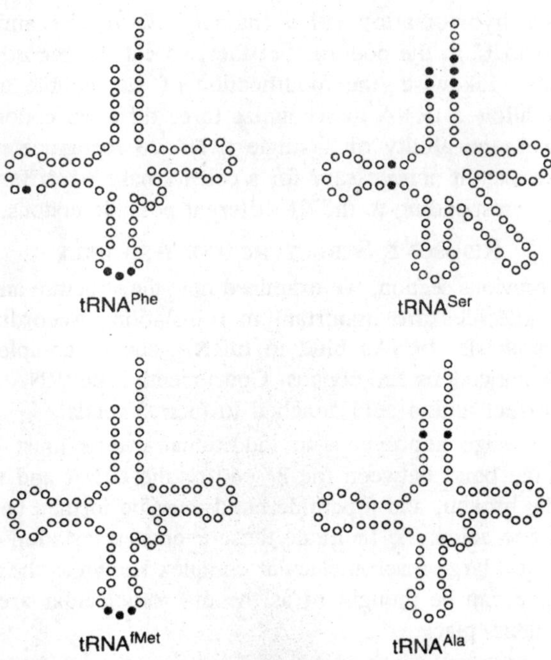

Fig. 5.3. Important recognition sites in E. coli tRNA molecules.

code. The genetic code is degenerate. This means that more than one codon can specify the same amino acid. With the exception of serine, arginine, and leucine, this degeneracy always occurs at the third position in the codon. For example, Valine is specified by GUU, GUC, GUA, and GUG. In all four cases, the first two bases are G and U. The third base, however, can be U, C, A, or G. To explain this pattern of degeneracy, Francis Crick proposed that it is due to "wobble" at the third position in the codon—anticodon recognition process. According to the *wobble hypothesis*, the first two positions pair strictly according to the A—U/G—C rule. However, the third position can tolerate certain types of mismatches. This proposal suggests that the base at the third position can actually move a bit so that hydrogen bonding can occur between the codon and anticodon.

Because of the wobble rules, a single type of tRNA can recognize more than one codon. For example, a tRNA with an anticodon sequence of 3'—AAG—5' can recognize a 5'—UUC—3' and a 5'—UUU—3' codon. The 5'—UUC—3' codon is a perfect match with this tRNA. The 5'—UUU—3' codon is mismatched according to the standard

RNA—RNA hybridization rules (namely, G in the anticodon is mismatched to U in the codon), but the two can fit according to the wobble rules. Likewise, the modification of the wobble base to an inosine can allow a tRNA to recognize three different codons. At the cellular level, the ability of a single tRNA to recognize more than one codon makes it unnecessary for a cell to make 61 different tRNA molecules corresponding to the 61 different possible codons.

Ribosome Structure and Assembly

In the previous section, we examined how the structure and function of tRNA molecules are important in translation. According to the adaptor hypothesis, tRNAs bind to mRNA due to complementarity between the anticodons and codons. Concurrently, the tRNA molecules have the correct amino acid attached to their 3' ends.

To synthesize a polypeptide, additional events must occur. In particular, the bond between the 3' end of the tRNA and the amino acid must be broken, and a peptide bond must be formed between the adjacent amino acids. To facilitate these events, translation occurs on the surface of a large macromolecular complex known as the ribosome. The ribosome can be thought of as the macromolecular arena where translation takes place.

In this section, we will begin by examining the biochemical compositions of ribosomes in prokaryotic and eukaryotic cells. We will then examine the key functional sites on ribosomes for the translation process.

Prokaryotic and Eukaryotic Ribosomes are Assembled from rRNA and Proteins

Prokaryotic cells have one type of ribosome, which is found within the bacterial cytoplasm. By comparison, eukaryotic cells are compartmentalized into cellular organelles that are bounded by membranes. Eukaryotic cells contain biochemically distinct ribosomes in different cellular locations. The most abundant type of ribosome functions in the cytosol, which is the region of the eukaryotic cell that is inside the plasma membrane but outside of the organelles. Besides the cytosolic ribosomes, all eukaryotic cells have ribosomes within an organelle known as the mitochondrion. In addition, plant cells have ribosomes in their chloroplasts. The compositions of mitochondrial and chloroplast ribosomes are quite different from that of the cytosolic ribosomes. Unless otherwise noted, the term "eukaryotic ribosome" refers to the ribosomes in the cytosol, not to those found within organelles. Likewise, the description of eukaryotic translation refers to translation via cytosolic ribosomes.

Each ribosome is composed of structures called the large and small subunits. This term is perhaps misleading, since each ribosomal subunit itself is formed from the assembly of many different proteins and rRNA molecules. In prokaryotic ribosomes, the 30S subunit is formed from the assembly of 21 different ribosomal proteins and one rRNA molecule; the 50S subunit contains 34 different proteins and two different rRNA molecules. The designations 30S and 50S refer to the rate that these subunits sediment when subjected to a centrifugal force. This rate is described as a sedimentation coefficient in Svedberg units (S), in honor of T. Svedberg, who invented the ultracentrifuge in the 1920s. Together, the 30S and 50S subunits form a 70S ribosome. In prokaryotes, the ribosomal proteins and rRNA molecules are synthesized in the cytoplasm, and the ribosomal subunits are assembled there.

By comparison, the synthesis of eukaryotic rRNA occurs within the nucleus, and the ribosomal proteins are made in the cytosol. The 40S subunit is composed of 33 proteins and an 18S rRNA; the 60S subunit is made up of 49 proteins and 5S, 5.8S, and 28S rRNAs The assembly of the rRNAs and ribosomal proteins to make the 40S and 60S subunits occurs within a specialized region of the nucleus known as the *nucleolus*. The 40S and 60S are then exported into the cytosol, where they associate to form an 80S ribosome during translation.

Ribosomes were first visualized by electron microscopy in the mid-1950s, and their protein synthesizing capability was demonstrated a few years later. The ribosome is a fairly complicated macromolecular structure. Certain regions of the ribosomal subunits have names that describe their geometrical shape. For example, a region on the 50S subunit is called the central protuberance, since it protrudes from the center of this sub unit. The large subunit also contains an indentation called the valley. Similarly, the 30S subunit has a cleft. In the 70S ribosome, the cleft and valley provide an opening through the ribosome. This opening is occupied by mRNA during the translation process.

Components of Ribosomal Subunits form Functional sites for Translation

To understand the structure and function of the ribosome at the molecular level, researchers must determine the locations and functional roles of the individual ribosomal proteins and rRNAs. In recent years, many advances have been made to ward a molecular understanding of ribosomes. This model shows the locations of many ribosomal proteins and rRNA molecules.

During prokaryotic translation, the mRNA lies in a space between the 30S and 50S subunits. As the polypeptide is being synthesized, it exits through a hole within the 50S subunit. Ribosomes also contain discrete sites where tRNA binds and the polypeptide is synthesized. In 1963, James Watson was the first to propose a two-site model for tRNA binding to the ribosome. These sites are known as the *peptidyl site* (*P site*) and aminoacyl site (*A site*). Twenty years later, Knud Nierhaus and Hans-Jorg Rheinberger at the Max Planck Institute for Molecular Genetics in Berlin proposed a three-site model. This model incorporated the observation that uncharged tRNA molecules can bind to a site on the ribosome that is distinct from the P and A sites. This third site is now known as the *exit site* (*E site*). In the next section, we will examine the roles of the A, P, and E sites during the three stages of translation.

Stages of Translation

Like transcription, the process of translation can be viewed as occurring in three stages: *initiation*, *elongation*, and *termination*. During initiation, the mRNA, initiator tRNA (i.e., the tRNA bound to the start codon), and ribosomal subunits assemble to form a complex. After the initiation complex is formed, the ribosome slides along the mRNA in the 5' to 3' direction, moving over the codons. As the ribosome moves, tRNA molecules sequentially bind to the mRNA and ribosome, bringing with them the appropriate amino acids. In this way, the amino acids are linked in the order dictated by the codon sequence in the mRNA. Finally, a stop codon is reached, signaling the end of translation. At this point, disassembly occurs and the newly made polypeptide is released. In this section, we will examine the components that are required for the translation process and consider their functional roles during the three stages of translation.

Many Components are Required for Translation to Occur

Protein synthesis is a vital cellular process. Much cellular energy and many proteins are devoted to translation. As already described in this chapter, there are three

components that are centrally important in the synthesis of polypeptides. An overview of their functions is described here:

1. *mRNA* is necessary to specify the amino acid sequence within a polypeptide. The sequence of codons within mRNA specifies the

sequence of amino acids within a polypeptide. In addition, the structure of the 5' end of the mRNA helps initiate translation at the correct start codon. A stop codon is also required to signal the end of translation.

2. *tRNA* molecules are necessary to decipher the genetic information coded within the mRNA. Due to the action of the aminoacyl tRNA synthetases, each tRNA carries with it the amino acid that corresponds to its anticodon. During translation, the anticodon in the tRNA binds to the codon in the mRNA.
3. *Ribosomes* are large macromolecular structures that act as the catalytic site for polypeptide synthesis. The ribosome allows the mRNA and tRNAs to be correctly positioned as the polypeptide is made.

Besides these components, many small molecules and proteins participate in the three stages of translation. For example, molecules such as ATP and GTP provide important sources of energy. These molecules are necessary because the molecular interactions between mRNA and tRNAs, and also the synthesis of the polypeptide chain, utilize a significant amount of energy. In addition, many cellular proteins function at specific stages of translation. The individual roles that many of these molecules and proteins play will also be described as we progress through the three stages of translation.

The Initiation Stage Involves the Binding of mRNA and the Initiator tRNA to the Ribosomal Subunits

Initiation, the first stage of translation, involves the binding of mRNA and the first tRNA to the ribosomal subunits. A specific tRNA functions as the initiator tRNA, which recognizes the start codon in the mRNA. In bacteria, the initiator tRNA carries a methionine that has been covalently modified to N-formylmethionine; this tRNA is called $tRNA^{fmet}$. In this modification, a formyl group (—CHO) is attached to the nitrogen atom in methionine after the methionine has been attached to the tRNA.

During translational initiation, the mRNA, $tRNA^{fmet}$ and ribosomal subunits associate with each other to form an initiation complex. The formation of this complex requires the participation of three initiation factors: IF1, IF2, and IF3. IF1 and IF3 are necessary to dissociate the 50S and 30S ribosomal sub units so that the 30S subunit can reinitiate with another mRNA molecule.

The binding of the mRNA to the 30S ribosomal subunit relies on the complementarity between a short region of the mRNA and rRNA within the 30S subunit. A sequence within bacterial mRNAs, known as the *ribosomal-binding site* or *Shine—Dalgarno sequence*, is involved in the binding of the mRNA to the 30S sub unit. The Shine—Dalgarno sequence within the mRNA is complementary to a short sequence within the 16S rRNA. This complementarity promotes the hydrogen bonding of the mRNA to the 30S subunit.

The binding of the $tRNA^{fmet}$ to the 30S subunit and the mRNA requires the function of IF2. In most cases, the $tRNA^{fmet}$ binds to the first start codon (namely, AUG), which is typically few nucleotides downstream from the Shine—Dalgarno sequence. The start codon is usually AUG, but in some cases it can be GUG or UUG. Even when the start codon is GUG (which normally encodes valine) or UUG (which normally encodes leucine), the first amino acid in the polypeptide is still a formylmethionine, because only a $tRNA^{fmet}$ can initiate translation. During translation, the formyl group or the entire formylmethionine may be removed from the polypeptide.

The initiation stage of prokaryotic translation is completed when the 50S ribosomal subunit associates with the 30S subunit. For this to occur, the initiation factors must be released from the 30S subunit.

In eukaryotes, the assembly of the initiation complex bears many similarities to that in prokaryotes. However, additional factors are required for the initiation process. Note that eukaryotic Initiation Factors are designated eIF to distinguish them from prokaryotic initiation factors. The initiator tRNA in eukaryotes carries methionine rather than formylmethionine (as in prokaryotes). A eukaryotic initiation factor, eIF2, binds directly to $tRNA^{met}$ to recruit it to the 40S subunit. Also, several initiation factors (CBPI [cap-binding protein], eIF4A, eIF4B, eIF4F, and others) bind to the mRNA. These initiation factors recognize the 7-methylguanosine cap structure, unwind any secondary structure in the mRNA so that it can bind to the ribosome, and aid in the identification of a start codon.

In eukaryotes, the identification of the correct AUG start codon differs greatly from that in prokaryotes. Most eukaryotic mRNAs contain a 7-methylguanosine cap structure at their 5' end. The recognition of this cap structure is key to the initial binding of mRNA to the ribosome. After this occurs, the next step is to locate an AUG start codon that is somewhere downstream from the 5' cap structure. In 1978, while at New York University, Marilyn Kozak proposed that the ribosome accomplishes this task by scanning along the mRNA in the 3' direction in search of an AUG start codon. In many, but not all, cases, the ribosome uses the first AUG codon that it encounters as a start codon. When a start codon is identified, the 60S subunit assembles with the aid of eIF5.

By analyzing the sequences of many eukaryotic mRNAs, researchers have found that not all AUG codons near the 5' end of mRNA can function as start codons. In some cases, the scanning ribosome passes over the first AUG codon and chooses an AUG farther down the mRNA. As it turns out, the sequence of nucleotides around the AUG codon plays an important role in determining whether or not it will be selected as the start codon by a scanning ribosome. The consensus sequence for optimal start codon recognition is shown here:

						Start Codon			
G	C	C	(A/G)	C	C	A	U	G	G
–6	–5	–4	–3	–2	–1	+1	+2	+3	+4

Aside from an AUG codon itself, a guanosine at the +4 position and a purine, preferably an adenine, at the –3 position are the most important for start codon selection. These rules for optimal translation initiation are called Kozak's rules. Presumably, this consensus sequence is recognized by a ribosomal component that facilitates the initiation of translation.

Polypeptide Synthesis Occurs During the Elongation Stage

During the elongation stage of translation, amino acids are added, one at a time, to the polypeptide chain. These steps are referred to as the *elongation cycle*. Even though this cycle is rather complex, it occurs at a remarkable rate. Under nor mal cellular conditions, a polypeptide chain can elongate at a rate of 15—18 amino acids per second in bacteria and 6 amino acids per second in eukaryotes!

In the elongation cycle, a tRNA brings a new amino acid to the ribosome so that it can be attached to the end of the growing polypeptide chain. At the upper left, a short polypeptide is attached to the tRNA located at the P site of the ribosome. A new tRNA carrying a single amino acid binds to the A site. This binding occurs because the anticodon in the tRNA is complementary to the codon in the mRNA.

The formation of a peptide bond between the amino acid at the A site and the growing polypeptide chain occurs via a *peptidyl transfer* reaction. This term means that the polypeptide is removed from the tRNA in the P site and transferred to the amino acid at the A site. This transfer is accompanied by the formation of a pep- tide bond between the amino acid at the A site and the polypeptide chain, lengthening the chain by one amino acid. The peptidyl transfer reaction is catalyzed by an enzyme complex known as *peptidyltransferase*, which is composed of several proteins and rRNA. Interestingly, researchers have speculated that the rRNA may act as a ribozyme to catalyze this chemical reaction.

After the peptidyl transfer reaction is complete, the ribosome moves, or translocates, to the next codon in the mRNA. This moves the tRNAs at the P and A sites to the E and P sites, respectively. Also note that the next codon in the mRNA is now exposed in the unoccupied A site. Finally, the uncharged tRNA exits the E site, and the cycle can begin all over again.

Termination Occurs When a Stop Codon is Reached in the mRNA

The final stage of translation, known as termination, occurs when a stop codon is reached in the mRNA. In most species, the three stop codons are UAG, UAA, and UGA. These codons are sometimes referred to as *amber* (UAG), *ocher* (UGA), and *opal* (UAA). (Note: The term amber, or brown stone, is the English translation of the name Bernstein, a graduate student at the California Institute of Technology who was involved in the discovery of the UAG codon. In keeping with this tradition, the lighthearted names ocher and opal were given to the other two stop codons.) The stop codons, also known as nonsense codons, are not recognized by a tRNA with a complementary sequence. Instead, they are recognized directly by proteins known as *release factors*. In bacteria, there are two different release factors, designated RF1 and RF2, that recognize stop codons. RF1 recognizes UAA and UAG, and RF2 recognizes UGA and UAA. A third release factor, RF3, is also required. In eukaryotes, a single release factor, eRF, recognizes all three stop codons.

At the top, the completed polypeptide chain is attached to a tRNA in the P site. A stop codon is located at the A site. In the next step, RF1 or RF2 binds to the stop codon at the A site and RF3 binds at a different location on the ribosome. After RF1 (or RF2) and RF3 have bound, the bond between the polypeptide and the tRNA is hydrolyzed. The polypeptide is then released from the ribosome. The final step in translational termination is marked by the disassembly of t RNA, ribosomal sub units, and the release factors.

A Polypeptide Chain has a Directionality from its Amino Terminus to its Carboxy Terminus

Polypeptide synthesis has a directionality that parallels the 5' to 3' orientation of the mRNA. During each cycle of elongation, a *peptide bond* is formed between the carboxyl group in the last amino acid of the polypeptide chain and the amino group in the amino acid being added. This occurs via a condensation reaction that releases a water molecule:

The first amino acid is said to be at the *N-terminal* or *amino terminal end* of the polypeptide. The term N-terminal refers to the presence of a nitrogen atom (N) at this end. By comparison, the last amino acid in a completed polypeptide is located at the *C-terminal* or *carboxyl terminal end.* A carboxyl group (COOH) is always found at this site in the polypeptide chain.

Bacterial Translation can Begin Before Transcription is Completed

While most of our knowledge concerning transcription and translation has come from genetic and biochemical studies, electron microscopy has also been an important tool in elucidating the mechanisms of transcription and translation. As described earlier in this chapter, electron microscopy (EM) has been a critical technique in facilitating our understanding of ribosome structure. In addition, EM can be used to visualize genetic processes such as translation.

The first success in the EM observation of gene expression was achieved in 1967 by Oscar Miller Jr. and his colleagues at the Oak Ridge National Laboratory in Tennessee. Prior to this experiment, biochemical and genetic studies had suggested that the translation of a bacterial structural gene begins before the mRNA transcript is completed. In other words, as soon as an mRNA strand is long enough, a ribosome will attach to the 5' end and begin translation—even before RNA polymerase has reached the transcriptional termination site within the gene. This is termed the coupling between transcription and translation in bacterial cells.

Only one of the DNA strands is used as a template for the synthesis of mRNA transcripts. Several different RNA polymerase enzymes have recognized this gene and begun to transcribe it. Since the transcripts on the right side are longer than those on the left, Miller concluded that transcription was proceeding from left to right in the micrograph. This EM image also shows the process of translation. Relatively small mRNA transcripts, near the left side of the figure, had a few ribosomes attached to them. As the transcripts became longer, additional ribosomes were attached to them. The term *polyribosome* or *polysome* is used to describe an mRNA transcript that has many bound ribosomes in the act of translation. In this early study, the nascent polypeptide chains were too small for researchers to observe. In later studies, as EM techniques became more refined, the polypeptide chains emerging from the ribosome were also visible.

When they were published, in 1970, these micrographs were remarkably similar to the diagrams present in genetics texts, illustrations that were based on biochemical and genetic studies of bacterial gene expression. These EM studies left little double that gene expression occurs via the transcription of DNA and the translation of mRNA.

The Sorting of Eukaryotic Proteins can Occur Post Translationally or Cotranslationally; Sorting Signals are Contained within the Amino Acid Sequence of a Protein

Unlike bacteria, eukaryotic cells cannot couple their transcription and translation, because the two processes occur in different cellular compartments. Transcription occurs in the nucleus, translation in the cytosol. The compartmentalization of eukaryotic cells also has other consequences for translation. Since they are compartmentalized into many different membrane-bound organelles, eukaryotic cells must sort their proteins into the correct compartment. In general, any particular protein is only meant to function in a single compartment. For example, the enzyme F_0F_1-ATP-synthetase functions in the mitochondrion and is not found in other locations in the cell. Cytoskeletal proteins such as tubulin and actin are found in the cytosol but not within the lumen of cellular organelles. To fulfill its function, each type of protein must be sorted into the correct cellular compartment. In eukaryotes, this sorting can occur during translation (termed *cotranslational sorting*) or after translation is completed (termed *posttranslational sorting*).

Most proteins begin their synthesis on ribosomes in the cytosol. Translation is completed in the cytosol for those proteins destined for the cytosol, nucleus, mitochondrion, chloroplast, or peroxisome. The uptake of proteins into the nucleus, mitochondrion, chloroplast, and peroxisome then occurs post translationally. By comparison, the synthesis of other eukaryotic proteins begins in the cytosol and then is temporarily halted until the ribosome has become bound to the membrane of the endoplasmic reticulum (ER). After the ribosome has bound to the ER membrane, translation resumes, and the polypeptide is synthesized into the ER lumen or ER membrane. This event is termed *cotranslational import*. Proteins that are destined for the ER, Golgi, lysosome, plasma membrane, or secretion are first directed to the ER via this type of cotranslational import mechanism. Genetics texts commonly focus attention on the importance of the amino acid

Table 5.1. Sorting signals in eukaryotic proteins

Type of signal	*Description*
Mitochondrial-targeting signal	Usually a short sequence at the amino terminus of a protein that contains several positively charged residues (i.e., lysine and arginine). This signal folds into an α-helix in which the positive charges are on one face of the helix.
Nuclear-localization signal	Can be located almost anywhere in the polypeptide sequence. The signal is four to eight amino acids in length and contains several positively charged residues and usually one or more prolines.
Peroxisomal targeting signal	A specific sequence of three amino acids, serine-lysine-leucine, which is usually located near the carboxyl terminus of the protein.
SRP signal (sorting to the ER)	A sequence of ~20 amino acids near the amino terminus that is composed of mostly nonpolar amino acids.
ER retention signal	A sequence of four amino acids, lysine-aspartic acid-glutamic acid-leucine, which is located at the carboxyl terminus of the protein.
Golgi retention signal	A sequence of 20 hydrophobic amino acids that forms a transmembrane domain flanked by positively charged residues.
Lysosomal-targeting signal	A patch of amino acids within the polypeptide sequence. Positively charged residues within this patch are though to play an important role. This patch causes lysosomal proteins to be covalently modified to contain a mannose-6-phosphate residue that directs the protein to the lysosome.
Destination of cellular protein	*Types of signal the protein contains within its amino acids sequence*
Cytosol	No signal required.
Mitochondrion	Mitochondrial-targeting signal.
Nucleus	Nuclear-localization signal.
Peroxisome	Peroxisomal-localization signal.
ER	SRP signal and an ER retention signal.
Golgi	SRP signal and a Golgi retention signal.
Lysosome	SRP signal and a lysosomal targeting signal.
Plasma membrane	SRP signal and the protein contains a hydrophobic transmembrane domain that anchors it in the membrane.
Secretion	SRP signal. No additional signal required.

sequence in dictating the structure and function of a protein. Even so, another vital function of the polypeptide sequence in eukaryotes is protein sorting. In eukaryotic proteins, there are short stretches of amino acid sequences that direct the protein to its correct cellular location. These short amino acid sequences are called *sorting* or *traffic signals*. Each traffic signal is recognized by specific cellular components that facilitate the sorting of the protein to its correct compartment, either cotranslationally or posttranslationally.

Concluding Remark

The translation of mRNA into a polypeptide sequence requires many cellular components, including *ribosomes*, tRNAs, mRNA, and small effector molecules. According to the *adaptor hypothesis*, the anticodons in tRNA molecules act as the translators of the genetic code within mRNA. Prior to translation, enzymes known as *aminoacyl-tRNA synthetases* recognize tRNAs and attach the correct amino acid to their 3' end. During translation, the anticodon in a tRNA binds to a codon in the mRNA due to their complementary sequences. Concurrently, the tRNA carries the correct amino acid, which is then added to the growing polypeptide chain. At the wobble position, the first base in the anticodon of the tRNA binds to the third base in the codon of the mRNA. In many cases, the wobble position can accommodate more than one type of base. In addition, modified bases in the tRNA can often be recognized by different bases in the m RNA.

The ribosomes are large macromolecular structures that provide a site for translation. Each ribosome is composed of one small and one large subunit; each subunit contains several proteins and rRNAs that assemble together. In eukaryotes, this assembly occurs within the *nucleolus*; the ribosomal subunits are then ex ported into the cytosol, where translation occurs.

The translation process occurs in three stages, known as *initiation*, *elongation*, and *termination*. During initiation, the ribosomal subunits, tRNA, and initiator tRNA assemble together. In prokaryotes, the *ribosomal-binding site* is located a few nucleotides upstream from the start codon. In eukaryotes, the ribosome binds near the 5' end of the mRNA and then scans along the mRNA until it finds a start codon. After the start codon has been identified, translation can proceed to the elongation phase, in which the polypeptide is made. During elongation, there are three sites on the ribosome for tRNA binding. An appropriate tRNA first binds to the A site when its anticodon recognizes the codon in the mRNA. A peptide bond then forms between the amino acid attached to the tRNA at the A site and the growing polypeptide chain attached to a tRNA in the *P site*. The tRNA at the P site can then exit the ribosome from the *E site*. Finally, during the termination stage, the stop codon in the mRNA is reached. At this point, termination factors bind to the ribosome and cause the disassembly of the ribosomal subunits, tRNA, mRNA, and completed polypeptide chain.

Some interesting differences are found between prokaryotic and eukaryotic translation due to cell compartmentalization. Since bacteria contain a single intra cellular compartment, translation of mRNA can begin before transcription is completed. This is referred to as coupling between transcription and translation. Coupling does not occur in eukaryotes, because the cells are compartmentalized; transcription takes place in the nucleus, and translation occurs in the cytosol. Compartmentalization also requires eukaryotic cells to sort their proteins into the correct locations. This occurs via the presence of sorting signals within the translated polypeptides. The targeting of cytosolic, nuclear, mitochondrial, chloroplast, and peroxisomal proteins occurs *posttranslationally*, whereas proteins destined for the ER, Golgi, lysosomes, plasma membrane, or secretion are initially targeted to the ER in a *cotranslational* manner.

Our molecular understanding of translation has come from several lines of experimentation, including biochemistry, genetics, and microscopy. The structure and composition of tRNAs, mRNAs, and ribosomes have been biochemically investigated for many decades. From such studies, we know the structures of tRNAs, mRNAs, and even the ribosomes, which are highly complex. Electron microscopy has aided our elucidation of ribosome structure and has also provided a method to demonstrate that transcription and translation are coupled in bacteria.

The steps of translation have also been investigated at the molecular level. By modifying the amino acid attached to tRNA, Chapeville showed that the anticodon sequence in the tRNA is the determining factor during translation. This experimental observation supported the adaptor hypothesis proposed by Crick. We have also learned a great deal about the translation process by examining genetic sequences. The genetic code in mRNA was unraveled by the experiments of Nirenberg and Ochoa. In addition, geneticists have identified the functional roles of sequences such as the Shine—Dalgarno sequence in prokaryotes, and Kozak's rules in eukaryotes, by analyzing mutations that interfere with translation.

6

GENE REGULATION IN BACTERIA

Chromosomes of bacteria, such as *Escherichia coli*, contain a few thousand different genes. Some of these genes are regulated; others are not. The term *gene regulation* means that the level of gene expression can vary under different conditions. By comparison, unregulated genes have essentially constant levels of expression in all conditions over time. These are called *constitutive genes*. Frequently, constitutive genes encode proteins that are always necessary for the survival of the bacterium. In contrast, the majority of genes are *regulated* so that the proteins they encode can be produced at the proper times and in the proper amounts. The benefit of regulating genes is that the encoded proteins will only be produced when they are required. Therefore, gene regulation allows a cell to avoid wasting its valuable cellular energy making proteins it does not need. From the viewpoint of natural selection, this enables a bacterium to compete as efficiently as possible for a limited amount of available resources. This is particularly important since bacteria find themselves in an environment that is frequently changing with regard to temperature, nutrients, and many other factors. A few common processes that are regulated at the genetic level are listed here:

1. *Metabolism*: Some proteins function in the metabolism of small molecules. For example, certain enzymes are needed for a bacterium to metabolize a particular sugar. These enzymes are only required when the bacterium is exposed to that sugar in its environment.
2. *Response to environmental stress*: Certain proteins help a bacterium to survive environmental stress (e.g., osmotic shock, heat shock).

These proteins are only required when the bacterium is confronted with the stress.

3. *Cell division*: Some proteins are needed for cell division. These are only necessary when the bacterial cell is getting ready to divide.

The expression of *structural genes*, which encode distinct polypeptides, ultimately leads to the production of functional cellular proteins. As we have seen, gene expression is a multistep process that proceeds from transcription to translation, and it may involve posttranslational effects on protein structure and function. Gene regulation can occur at any of these levels of gene expression. In this chapter, we will examine the molecular mechanisms that account for these types of gene regulation. We will also consider how gene regulation affects the phenotype of the bacterium.

Transcriptional Regulation

In bacteria, the most common way to regulate gene expression is by influencing the rate at which transcription is initiated. Although we frequently refer to genes as being "turned on or off," it is more accurate to say that the level of gene expression is increased or decreased. At the level of transcription, this means that the rate of RNA synthesis can be increased or decreased.

In most cases, transcriptional regulation involves the actions of regulatory proteins that can bind to the DNA and affect the rate of transcription of one or more nearby genes. There are two common types of regulatory proteins. A *repressor* is a regulatory protein that binds to the DNA and inhibits transcription, whereas *activators* are proteins that increase the rate of transcription. The term *negative control* refers to transcriptional regulation by repressor proteins, *positive control* to regulation by activator proteins.

In conjunction with regulatory proteins, small effector molecules often play a critical role in transcriptional regulation. However, small effector molecules do not bind directly to the DNA to alter transcription. Rather, an effector molecule exerts its effects by binding to a regulatory protein such as an activator or repressor. The binding of the effector molecule causes a conformational change in the regulatory protein and thereby influences whether or not the protein can bind to the DNA. Genetic regulatory proteins that respond to small effector molecules have two functional domains. One domain is a site where the protein binds to the DNA; the other domain is the binding site for the effector molecule.

Small effector molecules are given names that describe how they affect transcription when they are present. (The regulatory proteins are given names that de scribe how they affect transcription when they are *bound* to the DNA.) An *inducer* is a small molecule that will cause transcription to increase. An inducer may accomplish this in two ways: it could bind to an activator protein and cause it to bind to the DNA; or it could bind to a repressor protein and prevent it from binding to the DNA. In either case, the transcription rate is increased. Genes that are regulated in this manner are called *inducible*.

Alternatively, the presence of a small effector molecule may inhibit transcription. This can also occur in two ways. A *corepressor* is a small molecule that binds to a repressor protein, thereby causing the protein to bind to the DNA. An *inhibitor* binds to an activator protein and prevents it from binding to the DNA. Both core compressors and inhibitors act to reduce the rate of transcription. Therefore, the genes they regulate are *repressible*.

Unfortunately, this terminology can be confusing, since a repressible system could involve an activator protein or an inducible system could involve a repressor protein. In this section, we will examine several examples where genes are regulated by the actions of genetic regulatory proteins that influence the rate of transcription.

Phenomenon of Enzyme Adaptation is Due to the Synthesis of Cellular Proteins

To a significant extent, our understanding of molecular genetics can be traced back to the creative minds of Jacques Monod and Francois Jacob at the Pasteur Institute in France. Their research into genes and gene regulation stemmed from an interest in the phenomenon known as *enzyme adaptation*, which had been identified at the turn of the 20th century. Enzyme adaptation refers to the observation that a particular enzyme appears within a living cell only after the cell has been exposed to the substrate for that enzyme. When a bacterium is not exposed to a particular substance, it does not make the enzymes needed to metabolize that substance. However, when the bacterium is exposed to the substance, it will synthesize the enzymes that can metabolize it. In the 1950s, Jacob and Monod focused their attention on lactose metabolism in *E. coli* to investigate this problem. Some key experimental observations that led to an understanding of this genetic system are listed here:

1. The exposure of bacterial cells to lactose increased the levels of lactose-utilizing enzymes by 1000- to 10,000-fold.

2. Antibody and labeling techniques revealed that the increase in the activity of these enzymes was due to the increased synthesis of the enzymes.
3. The removal of lactose from the environment caused an abrupt termination in the synthesis of the enzymes.
4. Mutations that prevented the synthesis of particular enzymes involved with lactose utilization showed that each enzyme was encoded by a separate gene.

These critical observations indicated that enzyme adaptation is due to the synthesis of specific cellular proteins in response to lactose in the environment. In the following sections, we will learn how Jacob and Monod discovered that this phenomenon is due to the interactions between genetic regulatory proteins and small effector molecules. In other words, we will see that enzyme adaptation is due to the transcriptional regulation of genes.

The *Lac* operon Encodes Proteins that are Involved in Lactose Metabolism

In bacteria, it is common for a few structural genes to be arranged together in a regulatory unit that is under the transcriptional control of a single promoter. This arrangement is known as an *operon*. The biological advantage of an person organization is that it allows a bacterium to coordinately regulate a group of genes that encode proteins with a common function. The key feature of an person is that the expression of the structural genes occurs as a single unit.

An operon contains several different regions. For transcription to take place, an operon is flanked by a *promoter* to signal the beginning of transcription and a *terminator* to signal the end of transcription. Two or more structural genes are found between these two sequences. To control the ability of RNA polymerase to transcribe an operon, an additional DNA sequence, known as the *operator sites*, is present

The organization of the genes within the lactose operon in *E. coli*. There are actually two separate transcriptional units. The first unit, known as the *lac* operon, contains a promoter and three structural genes, *lacZ*, *lacY*, and *lacA*. *LacZ* encodes the enzyme β-galactosidase, which enzymatically cleaves lactose and lactose analogues. As a side reaction, β-galactosidase also converts a small amount of lactose into *allolactose*, a related sugar. As we will see later, allolactose acts as a small effector molecule to regulate the *lac* operon. The *lacY* gene encodes lactose permease, a membrane protein required for the active transport of lactose into the cytoplasm of the bacterium. The *lacA* gene encodes a transacetylase enzyme that can covalently modify lactose and lactose analogues. Although the functional necessity of the transacetylase remains unclear, the acetylation of non-metabolizable lactose analogues may prevent their toxic buildup within the bacterial cytoplasm.

In the vicinity of the *lac* promoter are two regulatory sites, designated the operator and the CAP site. The *operator* is a sequence of nucleotides that provides a binding site for a repressor protein, and the *CAP site* is a DNA sequence recognized by an activator protein called the *catabolite activator protein* (CAP).

A second transcriptional unit involved in genetic regulation is the *lacI* gene. The *lacI* gene encodes the *lac* repressor, a protein that is important for the regulation of the *lac* operon. The *lac* repressor functions as a tetramer, composed of four identical subunits. The *lacI* gene has its own promoter, the *i* promoter, which is constitutively expressed at fairly low levels. The amount of *lac* repressor made is approximately ten tetramer proteins per cell. Only a small amount of the *lac* repressor protein is needed to repress the *lac* operon.

The Lac operon is Regulated by a Repressor Protein

The *lac* operon can be transcriptionally regulated in more than one way. The first method that we will examine is an inducible, negative control mechanism. This form of regulation involves the *lac* repressor protein, which binds to the sequence of nucleotides found within the *lac* operator site. Once bound, the *lac* repressor prevents RNA polymerase from sliding part the operator site and transcribing the *lacZ*, *lacY*, and *lacA* genes.

The ability of the *lac* repressor to bind to the operator site depends on whether or not allolactose (a product of lactose) is bound to it. The repressor protein functions as a tetramer; each of the four subunits has a single binding site for the inducer (namely, allolactose). When allolactose binds to the repressor, a conformational change occurs that prevents the *lac* repressor from binding to the operator site. Under these conditions, RNA polymerase is now free to transcribe the operon. In genetic terms, we would say that the operon has been *induced*.

To better appreciate this form of regulation at the cellular level, let's consider the process as it occurs over time. The effects of external lactose on the regulation of the *lac* operon. In the absence of lactose, no inducer is available to bind to the *lac* repressor. Therefore, the *lac* repressor binds to the operator site and inhibits transcription. In reality, the repressor does not completely inhibit transcription, so that a very

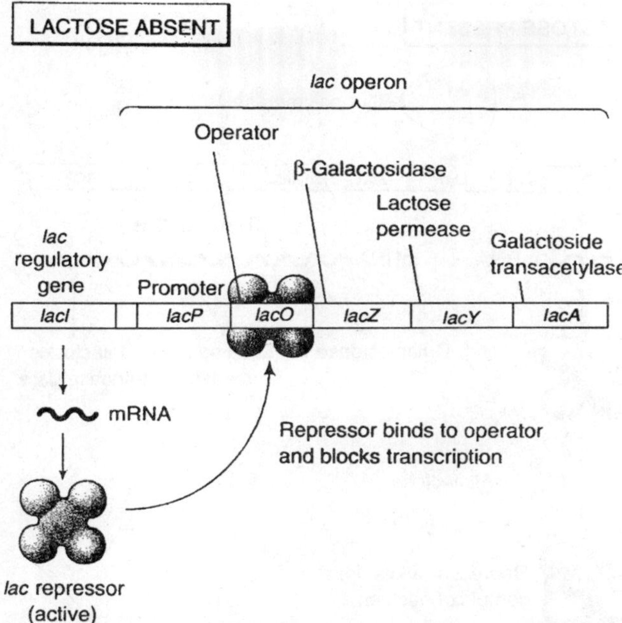

Fig. 6.1. Mechanism of induction of the lac operon. In the absence of the inducer allolactose, the repressor protein is tightly bound to the operator site, thereby inhibiting the ability of RNA polymerase to transcribe the operon.

small amount of β-galactosidase, lactose, permease, and transacetylase are made. However, the levels are too low for the bacterium to readily utilize lactose. When the bacterium is exposed to lactose, a small amount can be transported into the cytoplasm via the lactose permease, and β-galactosidase will convert some of it to allolactose. As this occurs, the cytoplasmic level of allolactose gradually rises; eventually, allolactose binds to the *lac* repressor. This prevents the repressor from binding to the *lac* operator site and thereby allows transcription of the *lacZ*, *lacY*, and *lacA* genes to occur.

To understand how the induction process is shut off in a lactose-depleted environment, let's consider the interaction between allolactose and the *lac* repressor. Allolactose does not bind irreversibly to the *lac* repressor. Rather, the *lac* repressor has a measurable affinity for binding and releasing allolactose. When the concentration of allolactose is above the affinity for the repressor protein, allolactose will bind to the *lac* repressor. In contrast, when the concentration of allolactose is below its affinity for the repressor, it will not be bound. With these

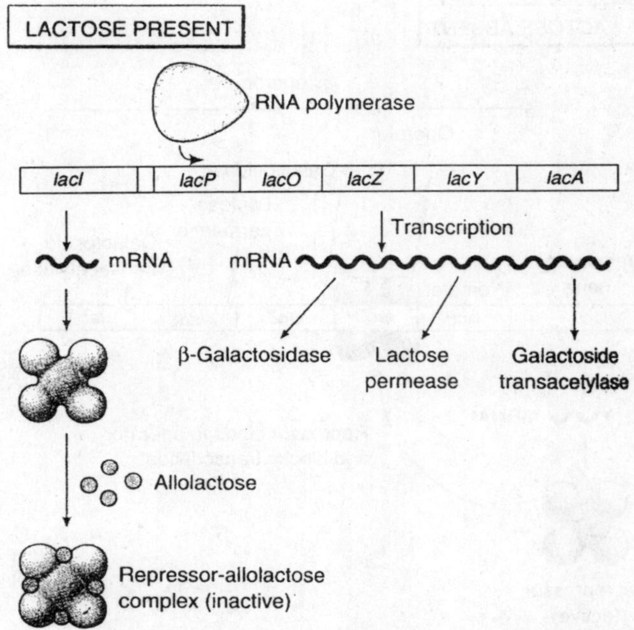

Fig. 6.2. Mechanism of induction of the lac operon. When there is sufficient allolactose, it binds to the response.

ideas in mind, we can see how the *lac* operon shut down when lactose is depleted from the environment. Since the cell is metabolizing its internal lactose, the cytoplasmic levels of lactose and allolactose will fall. Once the allolactose concentration within the cell drops below its affinity for the *lac* repressor, the unoccupied repressor will again bind to the *lac* operator site and cause repression. The proteins encoded by the *lac* operon have a finite half site, and they will eventually be degraded. This returns the cell to stage 1.

Jacob, Monod, and Pardee Investigated the Function of the lacI Gene

Now that we have an understanding of the *lac* operon, let's look back at one of the approaches that was used in the 1950s to elucidate the regulation of the *lac* operon. In their studies of lactose metabolism in bacteria, Jacob and Monod, along with their colleague, Arthur Pardee, had identified a few rare mutant strains of bacteria that had abnormal lactose adaptation. One type of mutant, which involved a defect in the *lacI* gene (designated *lacI*$^-$), resulted in the constitutive expression of the *lac* operon even in the absence of lactose. They were able to map genetically *lacI⁻* mutations using conjugation. The *lacI⁻* mutations mapped very close to the *lac* operon. From these observations, Jacob, Monod, and Pardee proposed two different possible functions of the *lacI* gene.

One hypothesis was that the *lacI* gene encodes a repressor protein that binds to the operon and inhibits transcription. (As we have already learned, this is the correct hypothesis.) Alternatively, since the *lacI* gene is physically next to the operon, a second possibility would be that the *lacI* gene is functioning as a site for the binding of a repressor protein produced by some other unknown gene. In other words, a second (incorrect) possibility would be that a *lacI* mutation prevents a repressor protein from binding to an operator site adjacent to the *lac* operon.

Jacob, Monod, and Pardee used a genetic approach to distinguish between these two alternative possibilities. Using bacterial conjugation methods, they introduced different portions of the *lac* operon into bacterial strains that carried *lacI* mutations. It is necessary to briefly review this process to understand this experiment. In general, bacteria are haploid and reproduce asexually by binary fission. On occasion, however, bacteria can mate with each other and transfer circular segments of DNA known as F factors. Sometimes, an F factor also carries genes that were originally found within the bacterial chromosome. These types of F factors are called F' factors.

In their studies, Jacob, Monod, and Pardee identified F' factors that carried portions of the *lac* operon. These F' factors can be transferred from donor to recipient strains during bacterial conjugation. For example, let's consider an F' factor that carries the *lacI* gene. A bacterium that has received this F' factor would actually have two copies of the *lacI* gene: one copy of the *lacI* gene on the chromo some and a second copy on the F' factor. A strain of bacteria containing F' factor genes is called a *merozygote*, or partial diploid.

The production of merozygotes was instrumental in allowing Jacob, Monod, and Pardee to elucidate the function of the *lacI* gene. There are two key points. First, the two *lacI* genes in a merozygote may be different alleles. For example, the *lacI* gene on the chromosome may be a defective *lacI⁻* allele while the *lacI* gene on the F' factor may be normal. Second, the genes on the F' factor and the genes on the bacterial chromosome are not physically adjacent to each other. According to the correct hypothesis, a *lacI* gene on an F' factor should be able to produce a repressor protein that could diffuse within the cell and eventually bind to the operator site of the *lac* operon located on the chromosome. In contrast, the incorrect hypothesis suggests that the *lacI* gene provides a binding site for a repressor protein. In this case, it must be physically adjacent to the *lac* operon to exert its effects. If the *lacI* gene functioned as a binding site for a repressor protein, a merozygote containing a normal *lacI* gene on an F' factor would not be able to affect the expression of the *lac* operon caused by a chromosomal mutation in the *lacI* gene. In this way, the approach of making merozygotes enabled Jacob, Monod, and Pardee to distinguish between these two possibilities.

Hypothesis

This experiment was done to determine whether the *lacI* gene codes for a protein that regulates the *lac* operon or whether the *lacI* gene serves as a repressor protein binding site and must be physically next to the *lac* operon to regulate it. The rationale is that if the *lacI* gene encodes a regulatory protein, then the *lacI* gene itself does not have to be physically next to the *lac* operon to repress it; the protein produced from the *lacI* gene (namely, the *lac* repressor) can diffuse throughout the cell and bind to an operator site regardless of the physical location of the *lacI* gene.

Testing the hypothesis

Starting material: The genotype of the recipient strain was $lacI^-$ $lacZ^+$ $lacY^+$ $lacA^+$. The donor strain had an F' factor that was $lacI^+$ $lacZ^+$ $lacY^+$ $lacA^+$

1. Mate the donor strain to the recipient strain. The mated strain contains a $lacI^-$ mutation on the chromosome and a normal *lacI* gene on the F' factor.
2. Grow an unmated recipient strain and a mated strain separately.
3. Divide each strain into two tubes.
4. In one of the two tubes, add lactose.
5. Incubate the cells long enough to allow *lac* operon induction.
6. Burst the cells with a sonicator. This allows β-galactosidase to escape from the cells.
7. Add β-*o*-nitrophenylgalactoside. This is a colorless compound. β-galactosidase will cleave the compound to produce galactose and *o*-nitrophenol (O-NP).O-nitrophenol has a yellow color. The deeper the yellow color, the more β-galactosidase was produced.
8. Incubate the sonicated cells to allow β-galactosidase time to cleave β-*o*-nitrophenylgalactoside (β-ONPG)
9. Measure the yellow color produced with a spectrophotometer.

Data

Strain	*Addition of lactose*	*Amount of β-galactosidase (percentage of parent strain)*
Recipient	No	100%
Recipient	Yes	100%
Mated	No	<1%
Mated	Yes	220%

Interpreting the data

In the recipient strain, the amount of yellow color produced was the same in the presence or absence of lactose. This is expected, since the expression of β-galactosidase in the *lacI*⁻ recipient strain was already known to be constitutive. In other words, the presence of lactose was no longer needed to induce the operon due to a defective *lacI* gene. In the mated strain, however, a different result was obtained. In the absence of lactose, the *lac* operons were repressed, even the operon on the chromosome. Since the normal *lacI* gene on the F' factor was not physically located next to the chromosomal *lac* operon, this result is consistent with the hypothesis that the *lacI* gene codes for a repressor protein that can diffuse throughout the cell and bind to any *lac* operon. In the presence of lactose, the operon is induced in the mated strain.

The interactions between regulatory proteins and DNA sequences illustrated in this experiment have led to the definition of two genetic terms. A *trans effect* is a form of genetic regulation that can occur even though two DNA segments are not physically adjacent. The action of the *lac* repressor is a trans effect. In contrast, a *cis effect* or a *cis-acting element* is a DNA segment that must be adjacent to the gene(s) that it regulates. The *lac* operator site is an example of a cis-acting element. A trans effect is mediated by genes that encode regulatory proteins, whereas a cis effect is mediated by DNA sequences that bind regulatory proteins.

The *lac* operon is also Regulated by an Activator Protein

The *lac* operon can be transcriptionally regulated in a second way, known as *catabolite repression*. This form of transcriptional regulation is influenced by the presence of the sugar glucose, which is a catabolite (it is broken down—i.e., catabolized—inside the cell). The presence of glucose ultimately leads to the repression of the *lac* operon. Glucose, however, is not itself the small effector molecule that binds directly to a genetic regulatory protein. Instead, this form of regulation involves a small effectors molecule, *cyclic AMP* (*cAMP*), which is produced from ATP via an enzyme known as *adenylate cyclase*. cAMP binds to an activator protein known as the *catabolite activator protein* (CAP) or the *cAMP receptor protein* (CRP).

The cAMP—CAP complex is an example of genetic regulation that is inducible and under positive control. When cAMP binds to CAR the cAMP—CAP complex binds to the CAP site near the *lac* promoter and increases the rate of transcription. The term catabolite repression may seem puzzling, since this region involves the action of an inducer and activator protein rather than a repressor. The term was coined before the action of the cAMP—CAP complex was understood. At that time, the primary observation was that glucose (a catabolite) inhibited (repressed) lactose metabolism. We now know that when a bacterium is exposed to glucose, the intracellular concentration of cAMP decreases because the presence of glucose inhibits adenylate cyclase, an enzyme needed for cAMP synthesis. When this happens, cAMP is no longer available to bind to the CAP. Therefore, the unoccupied CAP does not bind to the CAP site. This causes the transcription rate to decrease.

Catabolite repression enables the bacterium to utilize two sugars more efficiently When exposed to both glucose and lactose, *E. coli* cells will first utilize glucose, and catabolite repression will prevent the utilization of lactose. This is advantageous to the bacterium, because it does not have to express all the genes that are necessary for both glucose and lactose metabolism. If the glucose is used up, however, catabolite repression will be alleviated and the bacterium will then express the *lac* operon. The sequential use of two sugars by a bacterium is known as *diauxic growth*. It is a common phenomenon among many bacterial species. When glucose is one of the two sugars available, it is typical that the bacterium will metabolize glucose first, and then a second sugar after the depletion of glucose. Among *E. coli* and related species, diauxic growth is regulated by intracellular cAMP levels and the CAP.

In Recent Years, More Detailed Studies have Revealed that the *lac* operon has Three Operator Sites for the *lac* repressor

Our traditional view of the regulation of the *lac* operon has been modified as we have gained a greater molecular understanding of the process. In particular, detailed genetic and crystallographic studies have shown recently that the binding of the *lac* repressor is more complex than originally realized. The site in the *lac* operon that is commonly called the operator site was first identified by mutations that prevented *lac* repressor binding. These mutations, called *lacO*^{c} mutants, resulted in the constitutive expression of the *lac* operon even in strains that make a normal *lac* repressor protein. *LacO*^{c} mutations were found to be localized in the *lac* operator site (which is now known as O_1). This led to the simple view that a single operator site was bound by the *lac* repressor to inhibit transcription.

In the late 1970s and 1980s, however, two additional operator sites were identified. These sites are called O_2 and O_3. O_1 is the operator site that is next to the promoter. O_2 is located downstream (3' direction) in the *lacZ* coding sequence, and O_3 is located slightly upstream (5' direction) from the CAP-binding site. The O_2 and O_3 operators were initially called pseudo-operators, because the absence of either one of them only decreased repression by two- to threefold. However, studies by Benno Muller-Hill and his colleagues at the Institute for Genetics in Koln, Germany, revealed a surprising result. If both O_2 and O_3 are missing, repression is dramatically reduced even when O_1 is present. When O_1 is missing, repression is nearly abolished. This result supported the hypothesis that the *lac* repressor must bind to two out of three operators to cause maximal repression. According to this view, the *lac* repressor can bind to O_1 and O_2, or to O_1 and O_3, but not tc O_2 and O_3. Since these operator sites are a fair distance away from each other, it was proposed that the binding of the *lac* repressor to two operator sites causes the DNA to form a loop.

The proposal that the *lac* repressor binds to two operator sites has been con firmed by crystallographic studies. X-ray crystallography can provide a detailed picture of the molecular structure of macromolecules. To study protein—DNA interactions using crystallography, one must purify a large number of complexes that contain the DNA-binding protein and a fragment of DNA that is bound to the protein. These complexes are then subjected to conditions that will cause them to crystallize with each other, much like the formation of a NaCl crystal. When subjected to X-rays, a crystal will produce a pattern on an X-ray film. A computer analysis of the X-ray pattern can reveal information about the arrangement of atoms within the crystal.

In 1996, the *lac* repressor was crystallized by Mitchell Lewis and his colleagues at the University of Pennsylvania and the Oregon Health Sciences University. The crystal structure of the *lac* repressor has provided exciting insights into its mechanism of action. As mentioned earlier in this chapter, the *lac* repressor is a tetramer of four identical subunits. The crystal structure revealed that each dimer within the tetramer recognizes one operator site. The amino acid side chains in the protein interact directly with bases in the major groove of the DNA helix. This is how genetic regulatory proteins can recognize specific DNA sequences. Each dimer within the tetramer is recognizing a single operator site. The association of two dimers to form a tetramer forces the two operator sites close to each other. For this to occur, a loop must form in the DNA. The formation of this loop is expected to inhibit the ability of RNA polymerase to transcribe this region.

A particularly striking observation is that the binding of the cAMP—CAP complex to the DNA causes a 90° bend in the DNA structure. When the repressor is active (i.e., not bound to allolactose), the cAMP—CAP complex facilitates the binding of the *lac* repressor to the O_1 and O_3 sites. When the repressor is inactive, this bending also appears to be important in the ability of RNA polymerase to initiate transcription slightly downstream from the bend.

The *Ara* operon can be Regulated Positively or Negatively by the same Regulatory Protein

Now that we have considered the regulation of *lac* operon, let's compare its regulation with that of other genes in the bacterial chromosome. Another operon in *E. coli* involved in sugar metabolism is the *ara* (arabinose) operon. The sugar arabinose is a constituent of the cell walls of a few types of plants. The *ara* operon contains three structural genes, *araB*, *araA*, and *araD*, encoding three enzymes involved in arabinose metabolism. The actions of the three enzymes metabolize arabinose into D-xylulose-5-phosphate.

Like the *lac* operon, the *ara* operon contains a single promoter, designated P_{BAD}. This operon also contains a CAP site for the binding of the catabolite activator protein. The *araC* gene, which has its own promoter (P_C), is adjacent to the *ara* operon. *AraC* encodes a regulatory protein, called the *araC* protein, that can bind to operator sites designated *araI*, *araO*$_1$, and *araO*$_2$.

As we have seen with the *lac* operon, some regulatory proteins such as the lac repressor inhibit transcription whereas others such as CAP turn on transcription. The araC protein is rather interesting and unusual because it can act as either a negative or positive regulator of transcription, depending on whether or not arabinose is present. In the absence of arabinose, the araC protein binds to the *araI*, *araO*$_1$ and *araO*$_2$ operator sites. An araC protein dimer is bound to *araO*$_1$ while monomers are bound at *ara*O_2 and *araI*. The binding of the araC protein to the *ara*O_1 site activates the transcription of the *araC* gene. In other words, the araC protein is a positive regulator of the *araC* gene.

In contrast, the araC proteins bound at *ara*O_2 and *araI* repress the arabinose operon. The araC proteins at *ara*O_2 and *araI* can bind to each other by causing a loop in the DNA, as originally proposed in 1990 by Robert Schleif and his colleagues at Johns Hopkins University. This DNA loop prevents RNA polymerase from transcribing the ara operon. Therefore, in absence of arabinose, the *ara* operon is turned off. When arabinose is bound to the araC protein, the interaction between the araC proteins at the *ara*O_2 and *araI* sites is broken. This breaks the DNA loop. In addition, a second araC protein binds at the *araI* site. This araC dimer at the *araI* operator site activates transcription. This activation can occur in conjunction with the activation of the *ara* operon by CAP and cAMP if glucose levels are low. When the *ara* operon is activated, the bacterial cell can efficiently metabolize arabinose.

The *Trp* operon is Regulated by a Repressor Protein and also by Attenuation

The *trp* operon (pronounced "trip") encodes enzymes that are needed for the biosynthesis of the amino acid tryptophan. The *trpE*, *trpD*, *trpC*, *trpB*, and *trpA* genes encode enzymes involved in tryptophan biosynthesis. The *trpL* and *trpR* genes are involved in regulating the *trp* operon in two different ways. The *trpR* gene encodes the *trp* repressor protein. When tryptophan levels within the cell are very low, the *trp* repressor cannot bind to the operator site. Under these conditions, RNA polymerase transcribes the operon. In this way, the cell expresses the genes that encode the synthesis of tryptophan. When the tryptophan levels within the cell become high, tryptophan acts as a corepressor to bind to the *trp* repressor protein. This causes a conformational change in the *trp* repressor, which allows it to bind to the *trp* operator site. This inhibits the ability of RNA polymerase to transcribe the operon. Therefore, when there is a high level of tryptophan within the cell (i.e., when the cell does not need to make more tryptophan), the *trp* operon is turned off.

After the action of the *trp* repressor was elucidated, Charles Yanofsky and coworkers at Stanford made a few unexpected observations. Two *trp* operon mutations were identified in which a region including the *trpL* gene was missing from the operon. These mutations actually had higher levels of expression of the other genes in the *trp* operon. In addition, other mutant strains were found that lacked the *trp* repressor protein. Surprisingly, these mutant strains still could inhibit expression of the *trp* operon in the presence of tryptophan. As is often the case, unusual observations can lead people into interesting avenues of study. By pursuing this research further, Yanofsky discovered a second regulatory mechanism in the *trp* operon, called *attenuation*. This mechanism is mediated by the trpL gene.

Attenuation can occur only in bacteria that normally couple the processes of transcription and translation. During attenuation, transcription actually begins, but it is terminated before the entire mRNA is made. A segment of DNA, termed the *attenuator*, is important in facilitating this termination. When attenuation occurs, the mRNA encoding the *trp* operon is made as only a short piece that terminates shortly past the *trpL* region. Since this short mRNA does not encode the genes required for tryptophan biosynthesis, this mechanism can prevent the expression of these genes.

The segment of the *trp* operon immediately downstream from the operator site plays a critical role during attenuation. The first coding sequence in the *trp* operon is the *trpL* region, which encodes a short peptide called the Leader peptide. Two features are key in the attenuation mechanism. First, there are two tryptophan codons within the sequence that encodes the *trp* leader peptide. As we will see later, these two codons provide a way to sense whether or not there is sufficient tryptophan for the bacterium to translate its proteins. Second, the RNA that is transcribed from this region can form stem-loop structures. The type of stem-loop structure that forms underlies attenuation.

Different combinations of stem-loop structures are possible due to interactions among four sequences within the RNA transcript. Region 2 is complementary to region 1 and also to region 3 is complementary to region 2 as well as to region 4. Therefore, several stem-loop structures are possible. Even so, keep in mind that a particular segment of RNA can only participate in the formation of one stem-loop structure. For example, if region 2 forms a stem-loop with region lit cannot (at the same time) form a stem-loop with region 3. Alternatively, if region 2 forms a stem-loop with region 3, then region 3 cannot form a stem-loop with region 4. Though several stem-loop structures are possible, the 3—4 stem-loop structure is unique: it can act as a transcriptional terminator. The 3—4 stem-loop and the U-rich attenuator sequence act as an intrinsic terminator. Therefore, the formation of the 3—4 stem-loop causes RNA polymerase to pause and the U-rich sequence dissociates from the DNA. This terminates transcription at the 3' end of the *trpL* gene. By comparison, if region 3 forms a stem-loop with region 2, transcription will not be terminated because a 3-4 stem-loop cannot form.

Conditions that favor the formation of the 3—4 stem-loop ultimately rely on the translation of the *trpL* gene. There are three possible scenarios. Translation is not coupled with transcription. Since it is the most stable form of the mRNA, region 1 rapidly hydrogen bonds to region 2, and region 3 is left to hydrogen bond to region 4. Therefore, the terminator stem-loop forms, and transcription will be terminated just past the *trpL* gene. Coupled transcription and translation occur under conditions where the tryptophan concentration is low. When tryptophan levels are low, the cell cannot make a sufficient amount of $tRNA^{Trp}$. The ribosome pauses at the *trp* codons in the *trpL* gene, because it is waiting for the cell to provide $tRNA^{Trp}$. This pause occurs in such a way that the ribosome shields region 1 of the mRNA. This sterically prevents region 1 from hydrogen bonding to region 2. As an alternative, region 2 hydrogen bonds to region 3. Therefore, since region 3 is already hydrogen bonded to region 2, the 3—4 stem-loop structure cannot form. Under these conditions, transcriptional termination does not occur, and RNA polymerase transcribes the rest of the operon. This ultimately enables the bacterium to make more tryptophan.

Finally, coupled transcription and translation occur under conditions where there is sufficient tryptophan in the cell. In this case, translation of the *trpL* gene progresses to its stop codon, where the ribosome pauses. The pausing at the stop codon prevents region 2 from hydrogen bonding with any region and thereby enables region 3 to hydrogen bond with region 4. This terminates transcription. Of course, keep in mind that the *trpL* gene contains two tryptophan codons. For the ribosome to smoothly progress to the *trpL* stop codon, there must have been enough $tRNA^{Trp}$ in the cell to translate this gene. It follows that the bacterium must have a sufficient amount of tryptophan.

Since the cell does not need to synthesize more tryptophan, the rest of the transcription of the operon is terminated.

Repressible Operons Usually Encode Anabolic Enzymes, and Inducible Operons Encode Catabolic Enzymes

In the preceding parts of this chapter, we have seen that bacterial genes can be transcriptionally regulated in a positive or negative way—

often times both. The *lac* operon and arabinose operon are inducible systems regulated by sugar molecules that activate transcription of these operons. By comparison, the *trp* operon is a repressible operon regulated by a corepressor, tryptophan, that binds to the repressor and turns the operon off. In addition, an abundance of $tRNA^{Trp}$ in the cytoplasm can turn the operon off via attenuation.

In studying the genetic regulation of many operons, geneticists have noticed a general trend concerning inducible versus repressible regulation. 'When the genes in an operon encode proteins that function in the breakdown (i.e., catabolism) of a substance, they are usually regulated in an inducible manner. The substance to be broken down (or a related compound) often acts as the inducer. For example, allolactose and arabinose act as inducers of the *lac* and *ara* operons, respectively. An inducible form of regulation allows the bacterium to phenotypically express the appropriate genes only when they are needed to catabolize these sugars.

In contrast, other enzymes are important for synthesizing small molecules (i.e., anabolism). The genes that encode these anabolic enzymes tend to be regulated by a repressible mechanism. The inhibitor or corepressor is commonly the small molecule that is the product of the enzymes' biosynthetic activity. For example, tryptophan is produced by the sequential action of several enzymes that are en coded by the *trp* operon. Tryptophan it self acts as a corepressor that can bind to *trp* repressor protein when the intracellular levels of tryptophan become relatively high. This mechanism turns off the genes required for tryptophan biosynthesis when enough of this amino acid has been made. Therefore, genetic regulation via repression provides the bacterium with a way to prevent the overproduction of the product of an anabolic pathway.

Translational and Posttranslational Regulation

Though genetic regulation in bacteria is exercised predominantly at transcription, there are many examples where regulation is exerted at a later stage in gene expression. In some cases, specialized mechanisms have evolved to regulate the translation of certain mRNAs. The translation of mRNA occurs in three stages: initiation, elongation, and termination. Genetic regulation of translation is usually aimed at preventing the initiation step. In this section, we will examine a couple of ways that bacteria can regulate the initiation of translation.

The net result of translation is the synthesis of a protein. It is the activities of proteins within living cells that ultimately determine an organism's traits. There fore, any modification that alters protein function can be considered to affect gene expression. The term *posttranslational regulation* refers to the control of the functioning of proteins that are already present in the cell rather than regulation of transcription or translation. Posttranslational control can either activate or inhibit the function of a protein. Compared with transcriptional or translational control, posttranslational control can be relatively fast, occurring in a matter of seconds, which is an important advantage. In contrast, transcriptional and translational control typically require several minutes or even hours to take effect, since these two mechanisms involve the synthesis and turnover of mRNA and polypeptides. In this section, we will also examine ways that protein function can be regulated post translationally.

Repressor Proteins can Inhibit the Initiation of Translation

One common way to exert translational control involves the binding of a regulatory protein to the mRNA. A *translational regulatory protein* recognizes sequences within the mRNA, much as transcription factors recognize DNA sequences. In most cases, translational regulatory proteins act to inhibit translation. These are known as *translational repressors*. When a translational repressor protein binds to the mRNA, it can inhibit translational initiation in one of two ways. One possibility is that it can bind in the vicinity of the Shine--Dalgarno sequence and/or the start codon and thereby sterically block the ribosome's ability to initiate translation in this region. Alternatively, the repressor protein may bind outside of the Shine--Dalgarno/start codon region, but stabilize an mRNA secondary structure that prevents initiation.

The Synthesis of Antisense RNA can also Regulate Translation

A second way to regulate translation is via the synthesis of *antisense RNA*. This term refers to an RNA strand that is complementary to a strand of mRNA. The mRNA strand has the same sequence as the DNA sense strand (also known as the coding strand). To understand this form of genetic regulation, let's consider a trait known as osmoregulation, which is essential for the survival of most bacteria. Osmoregulation refers to the ability to control the amount of water inside the cell. Since the solute concentrations in the external environment may rapidly change between hypotonic and hypertonic conditions, bacteria must have an osmoregulation mechanism to maintain their internal cell volume. Otherwise, bacteria would be susceptible to the harmful effects of lysis or shrinking. In *E. coli*, an outer membrane protein encoded by the *ompF* gene is important in

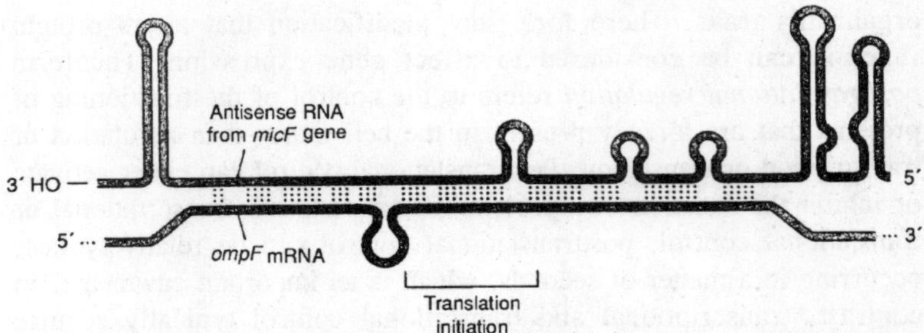

Fig. 6.3. The double-stranded RNA structure formed between the micF antisense RNA and the ompF mRNA.

osmoregulation. At low osmolarity, the OmpF protein is preferentially produced, whereas at high osmolarity its synthesis is decreased. The expression of an other gene, known as *micF*, is responsible for inhibiting the expression of the *ompF* gene at high osmolarity. The inhibition occurs because the *micF* RNA is complementary to the *ompF* mRNA. When the *micF* gene is transcribed, its RNA product binds to the *ompF* mRNA via hydrogen bonding between their complementary regions. The binding of the *micF* RNA to the ompF mRNA prevents the *ompF* mRNA from being translated. The RNA transcribed from the *micF* gene is called antisense RNA because it is complementary to the *ompF* RNA, which is a sense strand of mRNA that encodes a polypeptide. The *micF* RNA does not encode a polypeptide.

Posttranslational Regulation can occur via Feedback Inhibition and Covalent Modifications

A common mechanism to regulate the activity of metabolic enzymes is *feedback inhibition*. The synthesis of many cellular molecules such as amino acids, vitamins, and nucleotides occurs via the action of a series of enzymes that convert precursor molecules to a particular product. The final product in a metabolic pathway often can inhibit an enzyme that acts early in the pathway.

Enzyme 1 is an example of an *allosteric enzyme*. It contains two different binding sites. The *catalytic site* is responsible for the binding of the substrate and its conversion to intermediate A. The second site is a *regulatory site*. This site binds the product of the metabolic pathway. When bound to the regulatory site, the product inhibits the catalytic ability of enzyme 1.

To appreciate feedback inhibition at the cellular level, we can consider the relationship between the product concentration and the regulatory site on enzyme 1. As the product is made within the cell, its concentration will gradually increase. Once the product concentration has reached a level that is similar to its affinity for enzyme 1, it becomes likely that the product will bind to the regulatory site on enzyme 1 and inhibit its function. This blocks the further synthesis of intermediate A. In this way, the net result is that the product of a metabolic pathway inhibits the further synthesis of more product. Under these conditions, the concentration of the product has reached a level that is sufficient for the purpose of the bacterium.

A second strategy to control the function of proteins is by the covalent modification of their structure. Certain types of modifications are involved primarily in the assembly and construction of a functional protein. These alterations include proteolytic processing, disulfide bond formation, and the attachment of prosthetic groups, sugars, or lipids. These are typically irreversible changes required to produce a functional protein. In contrast, other types of modifications, such as phosphorylation ($-PO_4$) acetylation ($-COCH_3$) and methylation ($-CH_3$) are often reversible modifications that transiently affect the function of a protein.

Gene Regulation in the Bacteriophage Life Cycle

Viruses are small particles that contain genetic material surrounded by a protein coat. They can infect a living cell and then propagate themselves by utilizing the energy and metabolic machinery of the host cell. For this to occur, the genetic material of viruses orchestrates an intricate series of steps, involving the expression of many viral genes. During the past several decades, the reproduction of viruses has presented an interesting and challenging problem for geneticists to investigate. The study of bacteriophages, which are viruses that infect bacteria, has greatly advanced our basic knowledge of how genetic regulatory proteins work. In addition, the study of viruses has been instrumental in our ability to devise medical strategies aimed at combating viral disease. For example, our knowledge of the life cycles of human viruses has led to the development of drugs that are used to inhibit viral growth. For example, azidothymidine (AZT), which is used to combat HIV (human immunodeficiency virus), suppresses the production of viral DNA by inhibiting a viral gene product involved in viral DNA synthesis.

In this section, we will focus on the function of bacteriophage genes that encode genetic regulatory proteins. The structural genes of bacteriophages are often in an operon arrangement. This enables the genes within an operon to be controlled by regulatory proteins that bind to operator sites and influence the function of nearby promoters.

Like bacterial operons, phage operons can be con trolled by repressor proteins or activator proteins. To understand how this works, we will carefully examine the two life cycles of a virus called phage λ (lambda), which was discovered in the 1940s by Andre Lwoff and his colleagues at the Pasteur Institute in Paris. Since its discovery, phage λ has been investigated extensively and has provided geneticists with a model on which to base our understanding of viral proliferation.

Phage λ can Follow a Lytic or Lysogenic Life Cycle

Phage λ can bind to the surface of a bacterium and inject its genetic material into the bacterial cytoplasm. After this occurs, the phage will "choose" between two alternative life cycles, known as the lytic cycle and the lysogenic cycle. During the lytic cycle, the genetic instructions of the bacteriophage direct the synthesis of many copies of the phage genetic material and coat proteins. These are then assembled to make new phages. When synthesis and assembly is completed, the bacterial host cell is lysed, and the newly made phages are released into the environment.

Alternatively, phage λ can infect a bacterium and not progress to the lytic cycle. Instead, this phage can act as a *temperate phage* that will usually not produce new phages and will not kill the bacterial cell that acts as its host. Rather, the phage follows a lysogenic cycle. During the lysogenic cycle, phage λ integrates its genetic material into the chromosome of the bacterium. This integrated phage DNA is known as a *prophage*. A prophage can exist in a dormant state for a long time, during which no new bacteriophages are made. When a bacterium containing a lysogenic prophage divides to produce two daughter cells, it will copy the prophage's genetic material along with its own chromosome. Therefore, both daughter cells will inherit the prophage. At some later time, a prophage may become activated to excise itself from the bacterial chromosome and enter the lytic cycle. This process is called induction. During induction, the phage will promote the synthesis of new phages and, eventually, lyse the host cell.

Inside the virion head, phage λ DNA is linear. After injection into the bacterium, the two ends of the DNA become covalently attached to each other to form a circular piece of DNA. The organization of the genes within this circular structure reflects the two alternative life cycles of this virus. The genes in the top center of the figure are transcribed very soon after infection. This occurs at the beginning of either life cycle. As we will discuss later, the expression pattern of these early genes will determine whether the lytic or lysogenic cycle prevails. If the lysogenic cycle prevails, the integrase (*int*) gene will be subsequently turned on. The integrase gene encodes an enzyme that integrates the λ DNA into the bacterial chromosome.

If the lysogenic cycle is not chosen, the genes on the right side of the figure will be transcribed. This operon contains many genes that are necessary for λ assembly and release. These genes encode replication proteins, coat proteins, proteins involved in coat assembly, proteins involved in packaging the DNA into the phage head, and enzymes that will cause the bacterium to lyse.

During the Lytic Cycle, the λ Repressor and Integrase Proteins are Made

Now that we have an understanding of the phage λ life cycles and genome organization, let's examine how the decision is made between the lytic and lysogenic cycles. This choice depends on the actions of several genetic regulatory proteins. Our molecular understanding of the phage λ life cycles represents an extraordinary accomplishment in the field of genetic regulation. This process is quite detailed, since it involves a series of intricate steps in which regulatory proteins bind to several different sites in the λ genome. To simplify these events, we will begin by describing the steps that occur when the lysogenic cycle prevails. In the following section, we will examine the steps that occur when the lytic cycle is followed.

Soon after infection, two promoters, designated P_L and P_R are used for transcription. This initiates a competition between the lytic and lysogenic cycles. Initially, the transcription from P_R and P_L results in the synthesis of two short RNA transcripts that only encode two proteins called the *cro protein* and the *N protein*, both of which are genetic regulatory proteins. We will consider the function of the cro protein later in this chapter. The N protein is a genetic regulatory protein with an interesting function that we have not yet considered. Its function, known as *antitermination*, is to prevent transcriptional termination. The N protein binds to three sites, designated t_L, t_{R1}, and t_{R2} When it binds to the t_{R1} and t_{R2} sites, the transcript from P_R is extended to include *cII*, *O*, *P*, and *Q*. The *cII* gene encodes an activator protein, the O and P genes encode enzymes needed for the initiation of λ DNA synthesis, and the *Q* gene encodes a regulatory protein required for the lytic cycle. When the N protein binds to t_L the transcript from P_L is extended to include *int*, *xis*, and *cIII*. The *cIII* protein helps to stabilize the cII activator protein.

If the *cII—cIII* complex accumulates to sufficient levels, the lysogenic cycle is favored. Once it is made, the cII protein activates two different promoters in the λ genome. When the cII protein binds to the promoter P_{RE}, it turns on the transcription of *cI*, a gene that encodes the λ repressor. The cII protein also activates the *int* gene by binding to the promoter P_I. The λ repressor and integrase protein play central roles in promoting the lysogenic life cycle. When the λ repressor is made in sufficient quantities, it binds to operator sites that are adjacent to P_R and P_L. When the λ repressor is bound to these operator sites, it inhibits the expression of the genes that are required for lytic infection.

You may have noticed that the binding of the λ repressor to O_R will inhibit the expression of *cII*. This may seem counterintuitive since the cII protein was initially required to activate the *cI* gene (which en codes the λ repressor). You may be thinking that the inhibition of the *cII* gene will eventually prevent the expression of the λ repressor protein. This does not occur during the lysogenic cycle because the *cI* gene actually has two promoters, P_{RE} (which is activated by the cII protein) and also a second promoter called P_{RM}. Transcription from P_{RE} occurs early in the lysogenic cycle. P_{RE} gets its name because the use of this promoter results in the expression of the λ repressor during the establishment of the lysogenic cycle. The transcript made from the use of P_{RE} is very stable and quickly leads to a buildup of the λ repressor protein. This causes an abrupt inhibition of the lytic cycle because the binding of the λ repressor protein to O_R blocks the P_R promoter. Later in the lysogenic cycle, it is no longer necessary to make a large amount of the λ repressor. At this point, the use of the P_{RM} promoter is sufficient to make enough Repressor protein to Maintain the lysogenic cycle.

The Lytic Cycle Depends on the Action of the *Cro* Protein

As we have just seen, the λ repressor protein binds to O_R and prevents the expression of the operons needed for the lytic cycle. For the lytic cycle to occur, the λ repressor must be prevented from inhibiting P_R. This is the role of the cro protein. If the activity of the cro protein exceeds the activity of the cII protein, the lytic cycle prevails. As was mentioned, an early step in the expression of λ genes is the transcription from P_R to produce the cro protein. If the concentration of the cro protein builds to sufficient levels, it will bind to two operator regions, O_R and O_L. The binding of cro to O_L inhibits transcription from P_L; the binding of cro to O_R has several effects.

When the cro protein binds to O_R, it inhibits transcription from P_{RM} in the leftward direction. This inhibition prevents the expression of cI which encodes the λ repressor; the λ repressor is needed to maintain the progression to lysogeny. Therefore, the λ repressor cannot successfully shut down transcription from P_R.

The binding of the cro protein to O_R also allows a low level of transcription from P_R in the rightward direction. This enables the transcription and translation of the O, P, and Q genes. The O and P proteins are necessary for the replication of the λ DNA. The Q protein is a regulatory protein that activates another promoter, designated $P_{R'}$ (not to be confused with P_R). The $P_{R'}$ promoter controls a very large operon that encodes the proteins necessary for the phage coat, the assembly of the coat proteins, the packaging of the λ DNA, and the lysis of the bacterial cell. These proteins are made toward the end of the lytic cycle. The expression of these late genes leads to the synthesis and assembly of many new λ phages that are released from the bacterial cell when it lyses.

Cellular Proteases Influence the Choice between the Lytic and Lysogenic Cycle

As you may have noticed, the first two steps of the lysogenic and lytic cycles are identical. Whether the lysogenic or lytic cycle prevails depends on the steps that occur after the early genes are transcribed. In particular, the activity of the cII protein plays a key role in directing λ to the lysogenic or lytic cycle. A critical physiological issue is that the cII protein is easily degraded by cellular proteases that are produced by *E. coli*. Whether or not these proteases are made depends on the environmental conditions. If the growth conditions are very favorable (e.g., a rich growth medium), the intracellular protease levels are relatively high, and the cII protein tends to be degraded. When cII protein is present at a low level, it cannot bind to P_{RE} and activate the λ repressor gene which is required for the lysogenic cycle. Instead, the cro protein slowly accumulates to sufficient levels. The binding of the cro protein to O_R prevents transcription of the λ repressor gene from P_{RM} and, at the same time, allows the lytic cycle to proceed. In this way, environmental conditions that are favorable for growth promote the lytic cycle. This makes sense because a sufficient supply of nutrients is necessary to synthesize new bacteriophages.

Alternatively, starvation conditions favor the lysogenic cycle. When nutrients are limiting, cellular proteases are relatively inactive. Under these conditions, the cII protein will build up much more quickly than the cro protein. Therefore, the cII protein will turn on P_{RE} and lead to the transcription of the λ repressor. This event favors the lysogenic cycle. It is advantageous to favor lysogeny under starvation conditions, because there may not be sufficient nutrients available for the production of new λ phages.

After lysogeny has been established, certain environmental conditions will favor induction to the lytic cycle. For example, exposure to UV light will promote induction. This also is caused by the activation of cellular proteases. In this case, a cellular protein known as RecA (a protein ordinarily involved in facilitating recombination between DNA molecules) senses the DNA damage and is activated to become a cellular protease. RecA protein cleaves the λ repressor and thereby inactivates it. This allows transcription from P_R and eventually leads to the accumulation of the cro protein. This favors the lytic cycle. Under these conditions, it maybe advantageous for λ to make new phages and lyse the cell, because the exposure to UV light may have already damaged the bacterium and prevented further bacterial growth and division.

The O_R Region Provides a Genetic Switch between the Lytic and Lysogenic Cycles

Before we end this section on the λ life cycles, it is interesting to consider how the O_R region acts as a genetic switch between the lytic and lysogenic life cycles. De pending on the binding of genetic regulatory proteins to this region, the switch can be turned to favor the lytic or lysogenic cycle. To understand how this switch works, we need to take a closer look at the O_R region.

The O_R region contains three operator sites, designated O_{R1}, O_{R2}, and O_{R3}. The λ repressor protein or the cro protein can bind to any or all of these sites. The binding of these two proteins at these sites governs the switch between the lysogenic and lytic cycles. There are two critical issues that influence this binding event. The first is the relative affinities that the regulatory proteins have for these operator sites. The second is the concentrations of the λ repressor protein and the cro protein within the cell. Let's first consider how an increasing concentration of the λ repressor protein can switch on the lysogenic cycle and switch off the lytic cycle. This protein was first isolated by Mark Ptaslme and his colleagues while at Harvard University in 1967. Their studies showed that λ repressor binds with highest affinity to O_{R1}, followed by O_{R2} and O_{R3}. As the concentration of the λ repressor builds within the cell, a dimer of the λ repressor protein will first bind to O_{R1}, because it has the highest affinity for this site. Next, a second λ repressor dimer will bind to O_{R2}. This occurs very rapidly, because the binding of the first dimer to O_{R1} favors the binding of a second dimer to O_{R2}. This is called a *cooperative interaction*. The binding of the λ repressor to O_{R1} and O_{R2} inhibits transcription from P_R and thereby switches off the lytic cycle.

Early in the lysogenic cycle, the λ repressor protein concentration may become so high that it will occupy O_{R3}. Eventually, however, the λ repressor concentration will begin to drop, because the inhibition of P_R will decrease the synthesis of cII, which activates the λ repressor gene (from P_{RE}). As the λ repressor concentration gradually falls, it will first be removed from O_{R3}. This allows transcription from P_{RM}. In fact, the term λ repressor is somewhat misleading, because the binding of the λ repressor at only O_{R1} and O_{R2} acts as an activator of P_{RM}. The ability of the λ repressor to activate its own transcription allows the switch to the lysogenic cycle to be maintained.

In the lytic cycle, the binding of the cro protein controls the switch. The cro protein has its highest affinity for O_{R3} and has a similar affinity for O_{R2} and O_{R1}. Under conditions that favor the lytic cycle, the cro protein accumulates, and a cro dimer first binds to O_{R3}. This blocks transcription from P_{RM} and thereby switches off lysogeny. Later in the lytic cycle, the cro protein concentration will continue to rise, so that eventually it will bind to O_{R2} and O_{R1}. This turns down expression from P_R, which is not needed in the later stages of the lytic cycle. Genetic switches, like the one just described for phage λ represent an important form of genetic regulation. As we have just seen, a genetic switch can be used to control two alternative life cycles of a bacteriophage. In addition, genetic switches are also important in the developmental pathways for bacteria and eukaryotes. For example, certain species of bacteria can grow in a vegetative state when nutrients are abundant but will sporulate when conditions are unfavorable for growth. The choice between sporulation and vegetative growth involves genetic switches. Likewise, genetic switches operate in the developmental pathways in eukaryotes. Studies of the phage λ life cycle have provided fundamental information with which to understand how these other switches can operate at the molecular level.

Concluding Remark

Genes can be *regulated* in different ways. In prokaryotic cells, some genes are organized into multigene units called *operons*. An operon typically consists of a *promoter*, an *operator site*, several *structural genes*, and a terminator. The organization of an operon allows two or more genes to be coordinately regulated as a single unit. For transcriptional regulation, an important concept is that the binding of regulatory proteins to the operon can greatly influence the rate of transcription. Furthermore, the binding of small effector molecules to regulatory proteins can influence whether or not the regulatory proteins can bind to the DNA. Some regulatory proteins that exert *negative control*, such as the *lac* repressor and *trp* repressor, inhibit transcription. Others, such as the *catabolite activator protein* (CAP), exert *positive control* by enhancing transcription. The araC protein is interesting because it can act as both a *repressor* or an *activator* depending on the presence or absence of arabinose. Experiments involving the use of F factors showed that regulatory proteins are synthesized and can diffuse within the cell to ultimately bind to a distant operator site and cause repression. This action of a regulatory protein is called a *trans effect*. By comparison, operators are *cis regulators* by providing a binding site for genetic regulatory proteins.

Our understanding of transcriptional regulation has been greatly aided by studying the life cycles of viruses. In this chapter, we have considered the ability of phage λ to choose between the lytic and lysogenic cycle. This choice is determined by the actions of several genetic regulatory proteins that can bind to the λ DNA and influence the transcription of nearby genes. The O_R region provides a genetic switch for the lytic or lysogenic cycle. If the λ repressor controls the switch, the lysogenic cycle is favored, whereas binding of the cro protein to this switch favors the lytic cycle. Cellular proteases, which are influenced by environmental conditions, play a key role in the choice between the lytic and lysogenic cycles.

In addition to direct transcriptional regulation, gene expression in bacteria can be affected at later stages in the expression process. During *attenuation* of the *trp* operon, for example, transcription actually begins, but the length of the transcript is determined by the translation of the *trpL* gene. This form of regulation is unique to bacteria, because they couple the processes of transcription and translation. The translation of a complete mRNA transcript can also be regulated. One form of translational regulation involves the binding of *translational repressor* proteins that prevent translational initiation. A second way involves the binding of *antisense RNA* to the mRNA to block translation. Finally, the function of proteins that have already been made can be influenced *posttranslationally. Feedback inhibition* involves the non covalent binding of molecules to an *allosteric* site on a protein. This inhibition blocks the function of an enzyme that is required during an early step in a metabolic pathway. In addition, the function of proteins can be controlled by irreversible and reversible covalent modifications.

Overall, it is clear that genes can be regulated in a variety of different ways. If you are a scientist trying to understand the genetic regulation of a particular gene, your research can be a complicated, frustrating, and challenging task. At the same time, it can be a great deal of fun.

Early insights into the molecular mechanisms of gene expression came from experiments involving lactose metabolism in *E. coli*. Jacob and Monod examined the phenomenon of enzyme adaptation and concluded that the proteins involved in lactose metabolism were made after the bacterium was exposed to lactose. This observation led them to investigate the genes involved in this regulation process. By constructing merozygotes, they discovered that the *lacI* gene encodes a repressor protein that is bound to the operator site in the absence of lactose. More recently, studies by Muller-Hill showed that there are actually three operator sites. The crystal structure of the *lac* repressor tetramer bound at two out of three sites has been determined. Similarly, biophysical studies have determined the structure of CAP, a positive regulator of the *lac* operon.

A different form of genetic regulation, known as attenuation, was discovered by Yanofsky and colleagues. They identified mutant strains of bacteria that were missing the *trp* repressor, yet still could repress the *trp* operon when tryptophan levels were high. Additional mutations in the *trpL* region destroyed the ability to attenuate transcription. An analysis of the DNA sequence in this region revealed the ability of the mRNA to form alternative stem-loop structures. The 3—4 stem-loop structure forms a transcriptional terminator and thereby prevents the further synthesis of the *trp* operon.

An analysis of genetic sequences also revealed a posttranscriptional method of gene regulation involving antisense RNA. By comparing the sequence of the *ompF* mRNA and the *micF* RNA, researchers discovered that the two were complementary to each other. This explains how the synthesis of the *micF* RNA inhibits the translation of the *ompF* mRNA. When the (antisense) *micF* RNA hybridizes to the *ompF* mRNA, the latter cannot be recognized by the ribosomes. We concluded this chapter with a description of gene regulation in bacteriophage λ.

Many of the same kinds of experimental approaches were used to elucidate the molecular mechanisms that underlie the choice between the lysogenic and lytic cycles. In particular, many researchers have studied how mutations in particular genes in the λ genome favor the lysogenic or lytic cycles. In addition, biochemical and biophysical studies have analyzed the detailed interactions between λ regulatory proteins and the operator sites that they recognize.

7

Gene Regulation in Eukaryotes

Eukaryotic organisms, a category that includes protozoa, algae, fungi, plants, and animals, have many reasons for regulating their genes. Like their prokaryotic counterparts, eukaryotic cells need to adapt to changes in their environment. For example, eukaryotic cells can respond to changes in nutrient availability by enzyme adaptation much as prokaryotic cells do. Eukaryotic cells can also respond to environmental stresses such as UV radiation by inducing genes that provide protection against this harmful environmental agent. An example is the ability of sunbathers to develop a tan. The tanning response protects a person's cells against the damaging effects of ultraviolet rays.

Among plants and animals, multicellularity and a more complex cell structure also demand a much greater level of gene regulation. The life cycle of higher eukaryotic organisms involves the progression through several developmental stages to achieve a mature organism. Some genes are only expressed during early stages of development (e.g., the embryonic stage), whereas others are expressed in the adult. In addition, complex eukaryotic species are composed of many different tissues that contain a variety of cell types. Gene regulation is necessary to ensure the differences among distinct cell types.

It is amazing that the various cells within a multicellular organism usually contain the same genetic material, yet, phenotypically, may look quite different. For example, the appearance of a human nerve cell is about as similar to a muscle cell as an amoeba is to a paramecium. In spite of these phenotypic differences, a human nerve and muscle cell actually contain the same 46 chromosomes of the human genome. Nerve and muscle cells look so different because of gene regulation rather than differences in DNA content. There are many genes that are expressed in the nerve cell and not the muscle cell, and vice versa.

The molecular mechanisms that underlie gene regulation in eukaryotes bear many similarities to the ways that bacteria regulate their genes. Regulation of gene transcription is an important form of control. In addition, research in the past few decades has revealed that eukaryotic organisms frequently regulate gene expression at points other than transcription. In this chapter, we will discuss some well-studied examples where genes are regulated at many of these control points.

Eukaryotic species frequently regulate gene expression posttranslationally. For example, small effector molecules can regulate the activity of certain enzymes by feedback inhibition. In addition, the covalent posttranslational modification of proteins is a common way for eukaryotes to regulate the activities of cellular proteins.

Regulation of DNA and Chromatin Structure

Since genes are segments of DNA, it is not surprising that alterations in DNA structure can affect gene expression. This can occur in several ways. One possibility is that the number of copies of a gene can be increased. This is called *gene amplification*. Although this is an uncommon way to regulate gene expression, we will learn that it is occasionally used in eukaryotes. Another change in DNA that affects gene expression is the rearrangement of the DNA structure. *Gene rearrangement* occurs in the DNA of immune system cells to generate a diverse array of antibody proteins. A third, more common way to alter DNA structure is *DNA methylation*. In this section, we will examine how DNA methylation occurs and how it can regulate genes in a tissue-specific manner.

Since DNA is a component of chromosomes, the three-dimensional packing of chromatin is an important parameter affecting gene expression. If the chromatin is very tightly packed or in a *closed conformation*, transcription may be difficult or impossible. By comparison, chromatin that is in an *open conformation* is more easily accessible to transcription factors and RNA polymerase, so that transcription can take place. In this section, we will consider how conversion between the open and closed conformations can regulate the expression of genes.

The Number of Copies of a Gene can be Increased through DNA Amplification

While the majority of the genome remains constant in all of the cells of a eukaryotic organism, alterations in the amount of DNA occasionally occur during the normal course of eukaryotic development. For example, the oocytes of animals need to contain a large number of ribosomes to synthesize the vast amount of proteins that are stored in the egg. In the oocytes of the South African clawed toad (*Xenopus laevis*), for example, approximately 10^{12} ribosomes are required per oocyte. To construct so many ribosomes, a large amount of ribosomal RNA must be transcribed. To accomplish this task, the number of rRNA genes is increased through gene amplification. In this case, the mechanism involves an unusual DNA replication event in which the rRNA genes are replicated to form many copies of circular DNA molecules called *minichromosomes*. Each circular DNA molecule contains 1 to 20 copies of the rRNA gene. These minichromosomes accumulate within the nucleoli of the oocyte.

In laboratory cell lines, researchers can also select for the amplification of particular genes. For example, the drug methotrexate specifically inhibits an enzyme known as dihydrofolate reductase (DHFR). This enzyme, which catalyzes a step required in nucleotide biosynthesis, is essential for cell growth. Therefore, treating cells with methotrexate inhibits their growth. In a laboratory, cells can be exposed to this drug to select for cells that have become resistant to methotrexate inhibition. Methotrexate-resistant cells were determined to contain an increased number of genes that encode DHFR. This increase in the number of genes allows the cells to synthesize more DHFR and thereby overcome the inhibition from methotrexate. By analyzing the DNA of methotrexate-resistant cells, researchers have found that the increase in DHFR genes is due to circular minichromosomes or tandem repeats of the DHFR gene. In addition to these cases of gene amplification, a select number of genes are regulated by the rearrangement of DNA. A common example of this situation occurs during the development of the cells in the immune system. To generate a diverse array of antibodies to ward off infection, the cells of the immune system rearrange the DNA within the genes that encode antibodies. This occurs by a genetic recombination mechanism.

DNA Methylation Inhibits Gene Transcription

Another way that DNA structure can be modified is by the covalent attachment of methyl groups. DNA methylation is common in some eukaryotic species but certainly not all. For example, yeast and *Drosophila* have little or no detectable methylation of their DNA, whereas DNA methylation in vertebrates and plants is relatively abundant. In mammals, approximately 2 to 7% of the DNA is methylated. Eukaryotic DNA methylation occurs on cytosine residues at the number 5 position of the cytosine base. The sequence that is methylated is 5'-CG-3'. Note that this sequence also contains a cytosine in the opposite strand. A Then one strand is methylated, this is called *hemimethylation*, whereas the methylation of the cytosines in both strands is termed *full methylation*.

DNA methylation inhibits the transcription of eukaryotic genes, particularly when it occurs in the vicinity of the promoter. The methylation pattern may affect the ability of transcription factors and RNA polymerase to initiate transcription. In vertebrates, *housekeeping genes* typically encode proteins that are required in most cells of a multicellular organism. By comparison, other genes are highly regulated and may only be expressed in a particular cell type—*tissue-specific genes*—and/or at a particular stage of development. Many house keeping genes contain *CG islands* near their promoters. These CG islands, which are commonly 1000 to 2000 nucleotides in length, contain many unmethylated CG sequences. Therefore, housekeeping genes are not inhibited and tend to be expressed in most cell types. By comparison, many genes that are only expressed in certain cell types may also contain CG islands, but evidence is ac cumulating that DNA methylation may play an important role in the expression of these tissue-specific genes. The methylation of genes in particular tissues may prevent them from being expressed in the wrong cell type or at the wrong time.

DNA Methylation is Inheritable

Methylated DNA sequences are inherited during cell division. Experimentally, if fully methylated DNA is introduced into a vertebrate cell, the DNA will remain fully methylated even in subsequently produced daughter cells. However, if the same sequence of non-methylated DNA is introduced into a cell, it will remain non-methylated in the daughter cells. These observations indicate that the pattern of methylation is retained following DNA replication and, therefore, is inherited in daughter cells of future generations.

This model was originally suggested by Arthur Riggs at the City of Hope Research Institute in California and Robin Holliday and J. E. Pugh at the National Institute for Medical Research in London. In the parent cell, the DNA becomes methylated in both strands by *de novo* methylation (i.e., new methylation). When this cell divides, it will replicate its DNA and synthesize complementary daughter strands. Initially, the newly made daughter strands contain nonmethylated cytosines. This hemimethylated DNA is efficiently recognized by DNA methylase (also called DNA methyltransferase), which makes it fully methylated. This process is called *maintenance methylation*, since it preserves the methylated condition in future generations of cells. Overall, maintenance methylation appears to be an efficient process that routinely occurs within vertebrate and plant cells. By comparison, *de novo* methylation and demethylation are infrequent and highly regulated events. According to this view, the initial methylation or demethylation of a specific gene can be regulated at a specific time or stage of development. Once methylation has occurred, it can then be transmitted from parent to daughter cell via maintenance methylation.

As an example, let's consider DNA methylation within the context of vertebrate development. Suppose embryonic muscle cells have a gene that becomes fully methylated early in development due to *de novo* methylation. According to the model, this gene will be maintenance methylated in all of the muscle cells derived from these embryonic cells. Furthermore, since methylation inhibits transcription, all of these muscle cells will not transcribe this gene. In contrast, this same gene may not be methylated in embryonic nerve cells. There fore, it will be transcribed in the nerve cells of the organism. Along these lines, developmental geneticists are eager to determine how variations in DNA methylation patterns may be an important mechanism in the establishment of tissue characteristics during early stages of vertebrate development. Additional research will be necessary to understand how specific genes may be targeted for *de novo* methylation or demethylation during different developmental stages or in specific cell types.

Gene Accessibility can be Controlled by Changes in Chromatin Structure

Chromatin is composed of DNA and proteins that are organized into a compact structure that fits inside the nucleus of the cell. The DNA is wound around histone proteins to form nucleosomes that are 11 nm in diameter. This 11 nm fiber is condensed further to a 30 nm fiber, which is the predominant form of euchromatin found within the nucleus during interphase, when gene expression primarily occurs. Nevertheless, chromatin is a very dynamic structure that can alternate between highly condensed and highly extended conformations. The dynamic nature of chromatin is important in regulating gene transcription.

Variations in the degree of chromatin packing occur along the length of eukaryotic chromosomes during interphase. Tightly packed chromatin in a closed con formation cannot be transcribed. During gene activation, such chromatin must be converted to an open conformation that is less tightly packed than a 30 nm fiber. In certain cells, researchers can microscopically observe the decondensation of the 30 nm fiber when transcription is occurring. This chromosome does not form a uniform, compact 30 nm fiber. Instead, many decondensed loops radiate from the central axis of the chromosome. These loops are regions of DNA in which the genes are being actively transcribed. These chromosomes have been named *lampbrush chromosomes*, because their feathery appearance resembles the brushes that were once used to clean kerosene lamps.

Weintraub and Groudine used DNaseI Sensitivity to Study Changes in Chromatin Structure during Transcription of the β-globin Gene in 1976

To understand the interconversions between open and closed chromatin conformations, we need methods to evaluate the degree of chromatin packing as it occurs in living cells. An important method developed during the 1970s was the use of DNaseI to monitor DNA conformation. DNaseI is an endonuclease that cleaves DNA. DNaseI is much more likely to cleave DNA in an open conformation than that in a closed conformation, because a looser conformation allows greater accessibility of DNaseI to the DNA. Regions of DNA that are highly susceptible to DNaseI digestion are called *DNaseI hypersensitive sites*.

While working at Princeton University, Harold Weintraub and Mark Groudine used DNaseI sensitivity as a tool to evaluate differences in chromatin structure that occur when a gene is actively transcribed. As you may recall, humans possess several different genes that encode globin polypeptides. These polypeptides are the subunits of the oxygen-carrying protein hemoglobin. Prior to this work, it had been well established that globin genes are specifically expressed in red blood cells but not in other cell types such as brain cells and fibroblasts. Weintraub and Groudine asked the question, "Is there a difference in the chromatin packing of globin genes in cells that can actively transcribe the globin genes compared with that in cells in which the globin genes are turned off?" To answer this question, they used DNaseI sensitivity to compare the degree of globin gene packing in red blood cells versus that in brain cells and fibroblasts.

Since the globin genes are but a small part of the total chromosomal DNA, having a way to specifically monitor the digestion of the globin gene was a vital aspect of Weintraub and Groudine's experimental protocol. They accomplished this by using a cloned fragment of DNA (probe) that was complementary to the β-globin gene. This fragment was hybridized to the chromosomal DNA to determine specifically if the chromosomal globin gene was intact. If the chromosomal globin gene had been digested by DNaseI, it would not hybridize to the probe DNA because the corresponding chromosomal DNA would have been digested. However, if the chromosomal globin gene had not been digested by DNaseI, it could hybridize to the probe.

In Weintraub and Groudine's experiments, the cloned fragment of the globin gene was radiolabeled to allow detection of its presence. Following hybridization, the samples were then exposed to another enzyme, known as Si nuclease. This enzyme digests DNA strands but, interestingly, only cuts DNA when it is single- stranded, not when it is double-stranded. S1 nuclease would digest the radiolabeled DNA probe if it had not hybridized with the complementary chromosomal strand, but it would be unable to do so if the radiolabeled probe and chromosomal strand had formed a double-stranded structure.

After S1 digestion, Weintraub and Groudine reasoned that their sample would contain a radiolabeled DNA probe if the chromosomal DNA was in a closed con formation. This is because DNaseI would not digest the chromosomal gene, and therefore, the β-globin gene would be available to hybridize to the radiolabeled DNA probe. However, if the chromosomal globin gene was in an open conformation, DNaseI would digest the chromosomal DNA, preventing it from hybridizing with the radiolabeled DNA probe. In this situation, the radiolabeled DNA strand would be digested by S1 nuclease. Therefore, the susceptibility of the radiolabeled DNA to S1 nuclease digestion allowed them to evaluate whether the chromosomal globin gene was in an open or closed conformation.

Hypothesis

Changes in chromatin structure occur when globin genes are transcriptionally active.

Testing the hypothesis

Starting material: Nuclei were isolated from three different cell types in chicken: reticulocytes (immature red blood cells), brain cells, and fibroblasts. The globin genes are expressed in red blood cells but not in brain cells and fibroblasts.

1. Treat all three types of nuclei with the same amount of DNaseI.
2. Extract the DNA from the nuclei. This involves lysing the nuclei with detergent, removing the protein by treatment with a phenol—chloroform mixture, and then precipitating the DNA by adding ethanol. The precipitated DNA forms a pellet at the bottom of a test tube following centrifugation. The DNA at the bottom of the tube can then be resuspended in an appropriate solution for the next step.
3. Subject the DNA to sound waves (i.e., sonication) to break the DNA into fragments of an average length of 500 bp.
4. Add a radiolabeled DNA probe that is complementary to the β-globin gene.
5. Denature the DNA into single strands by treatment with high temperature. Then cool it to 65°C to allow the complementary DNA strands to hybridize with each other.
6. Reduce the temperature and divide each sample into two tubes. Into one tube of each sample add S1 nuclease. Omit S1 nuclease from the second tube.
7. Precipitate the double-stranded DNA with trichloroacetic acid.
8. Count the amount of radioactivity. (The technique of scintillation counting is described in the appendix.) The amount of radioactivity in the presence of S1 nuclease divided by the amount of radioactivity in the absence of S1 nuclease provides a measure of the percentage of radiolabeled DNA that has hybridized to the chromosomal DNA.

Data

Source of nuclei	% Hybridization of DNA probe
Reticulocytes	25%
Brain cells	>94%
Fibroblasts	>94%

Interrupting the data

In red blood cells, a much smaller percentage of the radiolabeled DNA probe hybridized to the chromosomal DNA than in brain cells and fibroblasts. These results were interpreted to mean that DNaseI is digesting the globin gene in the chromatin of red blood cells into small fragments that are too small to hybridize to the radiolabeled DNA probe. In other words, the globin genes in red blood cells are

DNaseI hypersensitive. By comparison, the globin genes in brain cells and fibroblasts are relatively resistant to DNaseI digestion. Since the globin genes are expressed in red blood cells but not in brain cells and fibroblasts, these results are consistent with the hypothesis that the globin gene is less tightly packed when it is being expressed. In cells where the globin genes should not be expressed (e.g., brain cells and fibroblasts), the chromatin containing the β-globin gene is tightly compacted. However, in red blood cells where this gene is expressed, the chromatin is more loosely packed so that transcription of the globin genes can occur. This phenomenon provides one way to regulate globin gene expression among different cell types.

Several Mechanisms have been Proposed for the Regulation of Chromatin Conformation

In recent years, geneticists have been trying to identify the proteins and DNA sequences that regulate the inter conversion between the closed and open conformations of chromatin. From a biochemical viewpoint, eukaryotic chromosomes are large, complex macromolecules. Therefore, it has been technically difficult to definitively identify all of the factors that influence chromatin structure within the nucleus of a living cell. Nevertheless, many different factors have been proposed to influence chromatin structure and thereby alter the transcriptional activity of genes.

Table 7.1. Factors that may regulate chromatin structure

Factor	*Effect*
Acetylation of core histones	The acetylation removes the positive charge on histones and may loosen the tight binding between DNA and the core histones.
Histone H1	The regulation of the linker histone may be involved in the interconversion of the 11 nm and 30 nm conformations.
High mobility group proteins	This broad category of DNA binding proteins may alter chromatin structure.
Transcription factors	Transcription factor proteins may influence the degree of chromatin packing or the position of nucleosomes.

One proposed mechanism for altering chromatin structure involves the post-translational modification of proteins that are bound to the DNA. For example, the modification of core histone proteins may be involved in chromatin decondensation. The histones within actively transcribed genes are covalently modified so that the lysine residues are acetylated. This acetylation eliminates the positive charge on the lysine residue and may loosen the tight interaction between the negatively charged DNA and the positively charged histone protein.

Besides acetylation, another suggested mechanism involves various proteins that may have important functions in altering chromatin structure during gene activation or repression. Histone H1, which is the linker histone, appears to play a critical role in the formation of the 30 nm fiber. Therefore, the regulation of histone H1 activity may contribute to the interconversion between the 30 nm diameter fiber and looser chromatin conformations. In addition, other nonhistone proteins may be important in regulating chromatin structure. A diverse group of DNA-binding proteins known as the *high mobility group proteins* are abundant within the nucleus of eukaryotic cells. While their role remains unclear, their relative abundance and ability to bind to DNA is consistent with the idea that they play an important role in chromatin structure.

Chromatin Packing and Nucleosome Location are Altered during Globin Gene Expression

Since the early studies of Weintraub and Groudine described in experiment, more detailed information has been gathered from molecular research in globin gene expression. Before discussing these studies, let's briefly consider globin gene organization and expression. Although the family of globin genes are expressed in red blood cells, individual members are expressed at different stages of development. For example, β-globin is expressed in adult red blood cells, whereas γ-globin is expressed in fetal red blood cells. Several of the globin genes are located adjacent to each other on the same chromosome.

Worldwide studies have been conducted that focus on inherited defects in hemoglobin composition. These disorders are known as *hemoglobinopathies*. An intriguing observation among certain patients is that they cannot synthesize β globin even though the DNA sequence of the β-globin gene is perfectly normal. This type of hemoglobinopathy has been found in Dutch, English, and Hispanic populations. It involves a DNA deletion that occurs upstream from the β-globin gene, although the β-globin gene itself is intact. Nevertheless, the β-globin gene is turned off in these patients. This unexpected finding prompted further investigations into how the globin genes are regulated.

Since the initial studies of hemoglobinopathies, a DNA region upstream from the β-globin gene has been identified as necessary for globin gene expression. This region, known as the *locus control region* (LCR), is involved in the regulation of chromatin opening and closing. It is missing in certain persons with hemoglobinopathies. As hypothesized, this DNA locus provides recognition sites for proteins that promote the general opening of the entire region. This gives RNA polymerase and gene-specific transcription factors access to the region. Not only does the locus control region affect the transcription of β-globin, it also influences the other globin genes in this region. Additional research will be necessary to determine if locus control regions are commonly involved with the regulation of chromatin packing for many eukaryotic genes.

Aside from the degree of chromatin packing, a second structural issue to consider is the position of nucleosomes. In chromatin, the nucleosomes are usually positioned at regular intervals along the DNA. The position of a nucleosome may greatly influence whether or not a gene can be transcribed. For example, if the TATA box is tightly bound to the histone core, it may be inaccessible to general transcription factors and RNA polymerase. Nucleosomes have been shown to change position in cells that normally express a particular gene but not in cells where the gene is inactive. For example, in fibroblasts that do not express the β-globin gene, nucleosomes are positioned at regular intervals from nucleotides –3000 to +1500. However, in red blood cells that can express the β-globin gene, a disruption in nucleosome positioning occurs in the region from nucleotide –500 to +200 This disruption maybe an important first step in gene activation.

Regulation of Transcription

In this text, we have frequently used the term transcription factor to describe proteins that influence the transcription of a given gene. In addition to the general transcription factors necessary for the basal level of transcription, eukaryotic cells also possess an interesting array of *regulatory transcription factors* that serve to regulate the transcription of nearby genes. As mentioned in the previous section, some proteins may affect transcription by altering the degree of chromatin packing or the position of nucleosomes. Alternatively, many regulatory transcription factors exert their effects by directly influencing RNA polymerase and/or general transcription factors.

Regulatory transcription factors commonly recognize a particular DNA sequence or consensus sequence. These sequences are analogous to the operator sites found near bacterial promoters. In eukaryotes, DNA sequences, recognized by regulatory transcription factors, are known as *response elements* or *control elements*. When a regulatory transcription factor binds to a response element, it affects the transcription of an associated gene. For example, the binding of regulatory transcription factors may facilitate the binding of general transcription factors and RNA polymerase in the vicinity of the promoter. This would enhance the rate of transcription. Alternatively, regulatory transcription factors may act as repressors by preventing transcription from occurring. In this section, we will examine several features of regulatory transcription factor function. We will begin by considering the structural features of transcription factor proteins and their response elements. We will then examine two well-studied examples that illustrate how the function of a transcription factor is modulated within a living cell.

Structural Features of Regulatory Transcription Factors Allow them to Bind to DNA

Genes that encode general and regulatory transcription factor proteins have been identified and sequenced from a wide variety of eukaryotic species including yeast, plants, and animals. Several different families of related transcription factors have been discovered. In recent years, the molecular structures of transcription factor proteins have become an area of intense research. Transcription factor proteins contain regions, called *domains*, that have specific functions. For example, one domain of a transcription factor may have a DNA-binding function, while another may pro vide a binding site for a small effector molecule. When a domain with a very similar structure or amino acid sequence is found in many different proteins, it is sometimes called a *motif*.

Several different domain structures found in transcription factor proteins. The protein secondary structure known as an α-helix is frequently important in the recognition of the DNA double helix. In helix-turn—helix and helix—loop—helix motifs, an α-helix called the *recognition helix* makes contact with and recognizes a base sequence along the major groove of the DNA. The major groove is a region of the DNA double helix where the bases contact the surrounding water in the cell. Hydrogen bonding between an α-helix and nucleotide bases is one way that a transcription factor can bind to a specific DNA sequence. Similarly, a zinc finger motif is composed of one α-helix and two β-sheet structures that are held together by a zinc (Zn^{++}) metal ion. The zinc finger also can recognize DNA sequences within the major groove. A second interesting feature of certain motifs is that they promote protein dimerization. The leucine zipper and helix—

loop—helix motifs mediate protein dimerization. The dimerization and DNA binding of two proteins that have leucine zippers. Alternating leucine residues in both proteins interact ("zip up"), resulting in protein dimerization. In some cases, two identical transcription factor proteins will come together to form a *homodimer* or two different transcription factors can form a *heterodimer*. As discussed later in this chapter, the dimerization of transcription factors can be an important way to modulate their function.

Regulatory Transcription Factors Recognize Response Elements that Function as Enhancers or Silencers

When the binding of a regulatory transcription factor to a response element in creases transcription, the response element is known as an *enhancer*. Enhancers can stimulate or *up-regulate* transcription 10- to 1000-fold. Alternatively, response elements that serve to inhibit transcription are called *silencers*. This is called *down-regulation*.

Many response elements are *orientation independent* or *bidirectional*. This means that the response element can function in the forward or reverse orientation. For example, if the forward orientation of an enhancer is

5'-GATA-3'
3'-CTAT-5,

then this enhancer will also bind to a regulatory transcription factor and enhance transcription even when it is oriented in the reverse direction:

5'-TATC-3'
3'-ATAG-5'

There is also striking variation in the location of response elements relative to a gene's promoter. In general, response elements are located in a region within a few hundred nucleotides upstream from the promoter site. Nevertheless, response elements are sometimes found several thousand nucleotides away from the actual promoter site. In some cases, response elements are located down stream from the promoter site and may even be found within introns! As you may imagine, the variation in response element orientation and location profoundly complicates the efforts of geneticists to identify the response elements that affect the expression of any given gene.

Different mechanisms have been proposed to explain how a regulatory transcription factor can bind to a response element and thereby affect gene transcription. Indeed, more than one mechanism may be involved. As mentioned, some regulatory transcription factors may influence transcription by altering chromatin packing or nucleosome positioning. This alteration can affect whether or not RNA polymerase can gain access to the promoter site. Alternatively, many regulatory transcription factors affect the ability of RNA polymerase and/or general transcription factors to function at the core promoter site. Since the response element may be a substantial distance away from the promoter site, these types of regulatory transcription factors can influence RNA polymerase by a looping mechanism.

According to this view, the regulatory transcription factor bound at the response element can physically influence events at the core promoter due to looping in the DNA. When the regulatory transcription factor enhances the rate of transcription, this is known as *transactivation*. The prefix trans- implies that the *activator* can bind to an enhancer site that may be a substantial distance away from the core promoter. As noted in a transactivating regulatory factor can exert this effect by recognizing and recruiting general transcription factors and RNA polymerase to the core promoter region.

The domain in an activator that stimulates transcription is called the *transactivating domain*. Most transactivating domains fall into one of three categories: acidic, glutamine rich, or proline-rich. Each category is characterized by the prevalence of a particular type of amino acid. For example, a yeast activator known as GAL4 has an acidic transactivating domain, which is 49 amino acids in length and contains 11 acidic (glutamic acid and aspartic acid) residues. As their name suggests, glutamine-rich and praline rich domains have a high proportion (typically 20—25%) of glutamine and proline, respectively.

Alternatively, other regulatory transcription factors may inhibit transcription by *transinhibition*. Transcription factors that cause transinhibition are also called *repressors*. They can inhibit transcription in a variety of ways. For example, repressors may physically interact with general transcription factors or RNA polymerase and prevent them from functioning. In addition, certain repressors exert their effects by inhibiting the function of transactivators.

The Function of Regulatory Transcription Factor Proteins can be Modulated in Three Ways

The function of regulatory transcription factor proteins can be affected directly in three common ways. These are (1) the binding of an effector molecule, (2) protein—protein interactions, and (3) covalent modification of the transcription factor itself. These three means of modulating regulatory transcription factor function. Usually, one or more of these modulating effects are important in determining whether a transcription factor can bind to the DNA and/or influence transcription by RNA polymerase. For example, an effector molecule may bind to a regulatory transcription factor and promote its binding to the DNA. Later in this chapter, we will see that steroid hormones function in this manner. An other important way is via protein—protein interactions. The formation of homodimers and heterodimers is a fairly common means of controlling transcription. Finally, the function of regulatory transcription factors can be affected by covalent modifications such as the attachment of a phosphate group. As discussed later, the phosphorylation of activators can alter their ability to stimulate transcription.

Steroid Hormones Exert their Effects by Binding to a Regulatory Transcription, Factor and Controlling the Transcription of nearby Genes

Thus far in this section, we have considered the general properties of transcription factor structure, the characteristics of the response elements they recognize, and the three ways that regulatory transcription factors can be modulated. We now will turn to specific examples that illustrate how regulatory transcription factors function within living cells. Our first example is a category that respond to *steroid hormones*. This type of regulatory transcription factor is known as a *steroid receptor*, because the steroid hormone binds directly to the protein.

The ultimate action of a steroid hormone is to affect gene transcription. Steroid hormones are synthesized by endocrine glands of animals and secreted into the bloodstream. The hormones are then taken up by cells that can respond to the hormones in different ways. For example, *glucocorticoid hormones* influence nutrient metabolism in most body cells by promoting glucose utilization, fat mobilization, and protein breakdown. Other steroid hormones, such as estrogen and testosterone, are called *gonadocorticoids* because they influence the growth and function of the gonads.

In this example, the hormone is transported into the cytosol of a cell. Once inside, there are glucocorticoid receptors that can specifically bind the hormone. Prior to hormone binding, the glucocorticoid receptor is complexed with proteins known as heat shock proteins, an example being Hsp90. After the hormone binds to the glucocorticoid receptor, Hsp90 is released. This exposes a nuclear localization signal within the receptor that allows it to travel into the nucleus through the nuclear pore. Two glucocorticoid receptors form a homodimer once inside the nucleus. The glucocorticoid receptor homodimer binds to glucocorticoid response elements (GREs) that are next to particular genes. The GREs function as enhancers. The binding of the glucocorticoid receptor homodimer to GREs activates the transcription of the adjacent gene, eventually leading to the synthesis of the encoded protein.

Animal cells usually have a large number of glucocorticoid receptors within the cytoplasm. Since GREs are located near several different genes, the uptake of many hormone molecules can activate many glucocorticoid receptors and thereby up- regulate many different genes. For example, if ten different genes within the nucleus contain GREs, all ten will be activated when the hormone is present. For this reason, a cell can respond to the presence of the hormone in a very complex way. Gluocorticoid hormones stimulate many genes that encode proteins involved in several different cellular processes, including the synthesis of glucose, the break down of proteins, and the mobilization of fats.

The CREB Protein is an Example of a Regulatory Transcription Factor Modulated by Protein—Protein Interaction and Covalent Modification

As we have just seen, some signaling molecules such as steroid hormones bind directly to regulatory transcription factors to alter their function. This enables a cell to respond to a hormone by up-regulating a particular set of genes. Most extracellular signaling molecules, however, do not enter the cell and bind directly to transcription factors. Instead, most signaling molecules must bind to receptors in the plasma membrane. This binding activates the receptor and leads the synthesis of an intracellular signal that causes a cellular response. One type of cellular response is to affect the transcription of particular genes within the cell.

As our second example of regulatory transcription factor function within living cells, we will examine the *cAMP response element—binding (CREB) protein*. The CREB protein is a regulatory transcription factor that becomes activated in response to specific cell-signaling molecules. It recognizes a response element with the consensus sequence 5'—TGACGTCA—3'. This response element, which is found near many different genes, has been termed a cAMP response element (GRE).

A wide variety of hormones, growth factors, neurotransmitters, and other signaling molecules can bind to plasma membrane receptors to initiate an intracellular response. In this case, the response involves the production of a second messenger, known as cAMP. (The extracellular signaling molecule itself is considered the primary messenger.) When the signaling molecule binds to the receptor, it activates a G protein that subsequently activates an enzyme known as adenylate cyclase. The activated adenylate cyclase catalyzes the synthesis of cAMP. The cAMP molecule then activates a second enzyme, protein kinase A. This enzyme can phosphorylate several different cellular proteins, including the CREB protein. When phosphorylated, the CREB protein stimulates transcription. In contrast, the unphosphorylated CREB protein can still hind to CREs but does not transactivate RNA polymerase.

Regulation of RNA Processing and Translation

In eukaryotic species, gene expression is commonly regulated at the RNA level. The function of RNA can be controlled in three general ways. One way is pre-mRNA processing. Following transcription, a pre-mRNA transcript is processed before it becomes a functional mRNA. We will see shortly how alternative splicing is regulated at the RNA level.

Another strategy for regulating gene expression is to influence the concentration of mRNA. As we have seen in this chapter, this can be done by regulating the rate of transcription. When the transcription of a gene is increased, a higher concentration of the corresponding RNA results. In addition, RNA concentration is greatly affected by the stability or half-life of a particular RNA. Factors that increase RNA stability are expected to raise the concentration of that RNA molecule. Later in this chapter, we will examine how sequences within mRNA molecules greatly affect their stability.

Finally, a third way to regulate RNA function is to control the ability of mRNAs to be translated. mRNA translation relies on the translational machinery (e.g., ribosomes, tRNA, etc.). The rate of translation can be regulated by influencing the functional activity of ribosomes or the ability of mRNA to be translated. In eukaryotes, one important strategy for regulating translation is to alter the rate of translation via the ribosomal machinery. This occurs primarily in two ways. One is to directly affect the function of translational initiation factors. This has the general effect of increasing or decreasing the translation of many mRNAs within the cell. Second, RNA-binding proteins can prevent ribosomes from initiating the translation process for specific mRNAs. In this section, we will examine both of these mechanisms for regulating mRNA translation.

Alternative Splicing Regulates which Exons Occur in an RNA Transcript, Allowing Different Proteins to be Made from the same Structural Gene

As already discussed the phenomenon of alternative splicing, although we did not consider how this process is regulated. During alternative splicing, a pre mRNA can be spliced in more than one way, leading to different combinations of exons in the resulting mRNAs. This produces two (or more) proteins that have specialized differences in structures and functions.

Alternative splicing is not a random event. Rather, the specific pattern of splicing is regulated in any given cell. The molecular mechanism for the regulation of alternative splicing is not entirely understood, although recent evidence indicates that it is due to variations in the concentrations of proteins known as *splicing factors*. Certain splicing factors play a key role in the choice of particular splice sites. One such category of splicing factors are the *SR proteins*. These splicing factors contain a domain at their carboxy terminal end that is rich in serine (S) and arginine (R). They also contain an RNA-binding domain at their amino terminal end.

In some cases, SR protein the ability of general splicing factors to choose the 5' splice site. The RNA-binding domain of the SR protein recognizes a region in the exon such as the 5' splice junction. It is hypothesized that the SR protein then recruits the general splicing factors into this region by using its SR (serine-arginine) domain. The net effect of SR protein binding and splicing factor recruitment is to promote exon recognition and inclusion of the recognized exon in the final mRNA product. Alternative splicing in different tissues may occur be cause each cell type has its own characteristic concentration of one or more types of SR proteins. These differences in SR protein concentration may play a key role in alternative splicing decisions.

The Stability of mRNA Influences mRNA Concentration

In eukaryotes, the stability of mRNAs can vary considerably. Certain mRNAs have very short half-lives (namely, several minutes), whereas others can persist for several days. In some cases, the stability of an mRNA can be regulated so that its half life is shortened or lengthened. A change in the stability of mRNA can greatly influence the cellular concentration of that mRNA molecule. In this way, factors that influence RNA stability can dramatically affect gene expression.

Various factors can play a role in mRNA stability. One important structural feature is the length of the polyA tail. Most newly made mRNAs contain a polyA tail that is approximately 200 nucleotides in length. This polyA tail is recognized by the *polyA-binding protein*. As an mRNA ages, its polyA tail tends to be shortened by the action of cellular exonucleases. Once it becomes less than 10 to 30 adenosines in length, the polyA-binding protein can not bind, and the mRNA is rapidly degraded by exo- and endonucleases.

Certain mRNAs, particularly those with short half-lives, contain sequences that act as destabilizing elements. While these destabilizing elements can be located anywhere within the mRNA, they are most commonly located at the 3' end between the stop codon and the polyA tail. This region of the mRNA is known as the *3'—untranslated region* (3'—UTR). An example of a destabilizing element is the *AU-rich element* (ARE) that is found in many short-lived mRNAs. This element, which contains the consensus sequence AUUUA, is recognized by cellular proteins that bind to the ARE and thereby influence whether or not the mRNA is rapidly degraded.

Phosphorylation of Ribosomal Initiation Factors can Alter the Rate of Translation

Modulation of translational initiation factors is widely used to control fundamental cellular processes. Under certain conditions, it is advantageous for a cell to stop synthesizing proteins. For example, if a cell is infected by a virus, it is vital to inhibit protein synthesis so that the virus cannot manufacture viral proteins. Likewise, if critical nutrients are in short supply it is beneficial for a cell to conserve its resources by inhibiting protein synthesis.

Several eukaryotic initiation factors are required to initiate translation. The phosphorylation of many different initiation factors has been found to affect translation. Two factors, eIF2 and eIF4F, appear to play a central role in controlling the initiation of translation. The functions of these two translational initiation factors are modulated by phosphorylation in opposite ways. When the α-subunit of eIF2 (known as eIF2α) is phosphorylated, translation is inhibited, whereas the phosphorylation of eIF4F increases the rate of translation.

A variety of conditions can lead to a shutdown of protein synthesis, including viral infection, nutrient deprivation, heat shock, and the presence of toxic heavy metals. These conditions promote the activation of protein kinases known as *eIF2α protein kinases*. Several eIF2α protein kinases have been identified. Once activated, eIF2α protein kinase can phosphorylate eIF2α. The phosphorylation of eIF2α causes it to bind tightly to another initiation factor subunit, known as eIF2B. Functional eIF2B is necessary so that eIF2 can promote the binding of the initiator $tRNA^{Met}$ to the 40S subunit However, when the phosphorylated eIF2α binds to eIF2B, it prevents eIF2B from functioning. Therefore, the initiator $tRNA^{Met}$ does not bind to the 40S subunit, and translation is inhibited.

A second important way to control translation is via the eIF4F translation factor that modulates the binding of mRNA to the ribosomal initiation complex. The function of eIF4F is stimulated by phosphorylation. A variety of conditions have been shown to cause eIF4F to become phosphorylated. These include the presence of growth factors, insulin, and other signaling molecules that promote cell proliferation. Conversely, conditions such as heat shock and viral infection decrease the level of eIF4F phosphorylation and thereby inhibit translation.

The Regulation of Iron Assimilation is an Example of the Regulatory Effect of RNA-binding Proteins on Translation

As we have just seen, the phosphorylation of ribosomal proteins can modulate the translation of mRNA. Since the ribosomes are necessary to translate all of a cell's mRNA, this form of regulation affects many mRNAs. By comparison, particular mRNAs are sometimes regulated by RNA-binding proteins that directly affect translational initiation or RNA stability. The regulation of iron assimilation provides a well-studied example in which both of these phenomena occur. Before discussing the translational control, it is interesting to consider the biology of iron metabolism.

Iron is an essential element for the survival of living organisms, since it is required for the function of many different enzymes. Iron ingested by an animal is absorbed into the bloodstream and becomes bound to a protein known as *transferrin*, a carrier of iron through the bloodstream. The transferrin—Fe^{++} complex is recognized by a *transferrin receptor* on the surface of cells; the complex binds to the receptor and then is transported into the cytosol by endocytosis. Once inside, the iron is then released from transferrin. At this stage, it may bind to cellular enzymes that require iron for their activity. Alternatively, if there is an overabundance of iron, the excess iron is stored within a hollow, spherical protein known as *ferritin*. The storage of excess iron within ferritin prevents the toxic buildup of too much iron within the cell.

Since iron is a vital yet potentially toxic substance, mammalian cells have evolved an interesting way to regulate iron assimilation.

The two mRNAs that encode ferritin and the transferrin receptor are both influenced by an RNA-binding protein known as the *iron regulatory protein* (IRP). This protein binds to a regulatory element within the mRNA known as the *iron regulatory element* (IRE). The ferritin mRNA has an IRE in its 5'—UTR. When IRP binds to this IRE, it inhibits the translation of the ferritin mRNA. However, when iron is abundant in the cytosol, the iron binds directly to IRP and prevents it from binding to the IRE. Under these conditions, the ferritin mRNA is translated to make more ferritin protein. The synthesis of ferritin prevents the toxic buildup of iron within the cytosol.

The transferrin receptor mRNA also contains iron response elements. How ever, the IREs in the transferrin receptor mRNA are located in the 3'—UTR. When IRP binds to these IREs, it does not inhibit translation. Instead, it increases the stability of the mRNA by blocking the action of RNA-degrading enzymes. This leads to increased amounts of transferrin receptor mRNA within the cell when the cytosolic levels of iron are very low. Under these conditions, more transferrin receptor is made. This promotes the uptake of iron when in short sup ply. In contrast, when iron is abundant within the cytosol, IRP is removed from the transferrin receptor mRNA, and the mRNA becomes rapidly degraded. This leads to a decrease in the amount of transferrin receptor and thereby helps to prevent the assimilation of too much iron into the cell.

Concluding Remark

In this chapter, we have surveyed a wide variety of mechanisms that allow eukaryotes to regulate gene expression. The DNA itself can be altered by *gene amplification*, *rearrangement*, or *methylation*. DNA methylation in vertebrates and higher plants appears to be an important mechanism to turn genes off in a *tissue-specific* or developmentally specific manner. In addition, the degree of chromatin packing is an important parameter that affects gene expression. For transcription to occur, the chromatin cannot be in a *closed conformation*. Instead, it must be in a loosely packed or *open conformation*. In the case of globin genes, a segment of DNA called the *locus control region* plays a role in regulating the packing of DNA in this region.

A wide array of *regulatory transcription factors* have been identified. These vary in their structures and modes of action. Some transcription factors bind to DNA and alter chromatin packing or nucleosome positioning. Other transcription factors recognize *response elements* in the vicinity of genes and influence the events that occur at the core promoter. *Enhancers* are response elements that *up-regulate* gene expression, whereas *silencers* have the opposite effect of *down-regulating* transcription. Response elements can be *bidirectional* and function at a fairly large distance away from the core promoter site. In this chapter, we have considered the molecular mechanism of two regulatory transcription factors that respond to cell hormones. The glucocorticoid receptor is a regulatory transcription factor that binds glucocorticoid hormone directly. Once the hormone is bound, it activates several cellular genes by binding to glucocorticoid response elements (GREs) that are next to genes. The CREB protein responds to intracellular levels of cAMP. The *CREB protein* binds to response elements known as CREs. When CREB becomes phosphorylated, it stimulates transcription by *transactivation*.

Gene regulation can also occur at the RNA level. For example, an RNA transcript can be regulated by alternative RNA splicing. This RNA processing event influences the type of protein that is made from the mRNA transcript. The stability of RNA can also be affected by particular sequences within the mRNA. Factors that promote RNA stability lead to a higher concentration of that RNA transcript. In addition, the rate of mRNA translation can be controlled in two ways. First, the translational machinery can be regulated by affecting the activity of translational initiation factors. Two particular factors, eIF2a and eIF4F, are modulated by phosphorylation in opposite ways. When eIF2cz is phosphorylated, translation is inhibited, whereas the phosphorylation of eIF4F increases the rate of translation. Second, specific RNA-binding proteins can affect the ability of certain mRNAs to be translated.

The regulation of gene expression in eukaryotes has been studied via many different techniques. In this chapter, we have considered a few approaches to studying this phenomenon. We saw, in one example, that Weintraub and Groudine were able to detect changes in the degree of DNA packing of the globin genes by using a DNaseI sensitivity assay. This approach exploits the fact that tightly packed chromatin is less susceptible to DNaseI digestion than is DNA in an open conformation. Furthermore, the identification of people who possess intact β-globin genes yet have hemoglobinopathies has suggested that a region near globin genes, known as the locus control region, is involved in controlling the degree of chromatin packing.

Other studies have focused on the identification of regulatory transcription factors that influence the expression of particular genes.

By comparing the structures of many different transcription factors, researchers have found that they tend to have common domains that act to either bind DNA or interact with effector molecules. The sequences of many response elements have also been determined. In a few cases, the combined efforts of many research groups have elucidated the detailed molecular mechanisms for the regulation of particular genes. For example, in this chapter, we have examined how the glucocorticoid receptor and CREB protein regulate genes at the level of transcription. We have also considered how IRP regulates ferritin and transferrin receptor expression at the level of translation.

8

GENE MUTATION AND REPAIR

As we have seen throughout this text, the function of DNA is to store the information for the synthesis of cellular proteins. A key aspect of the gene expression process is that the DNA itself normally does not change. This allows DNA to function as a permanent storage unit. However, on relatively rare occasions, a *mutation* can occur. The term mutation refers to a heritable change in the genetic material. This means that the structure of DNA has been changed permanently and this alteration can be inherited by daughter cells from a mother cell following cell division. If a mutation occurs in the germ line cells that produce gametes, it may also be passed from parent to offspring.

The topic of mutation is centrally important in all fields of genetics, including molecular genetics, Mendelian inheritance, and population genetics. Mutations provide the allelic variation that we have discussed throughout this text. For ex ample, phenotypic differences such as tall versus dwarf pea plants are due to mutations that alter the expression of particular genes. Mutations can have both beneficial and detrimental effects. On the positive side, mutations are essential to the continuity of life. They provide the variation that enables species to change and adapt to their environments. On the negative side, however, new mutations are much more likely to be harmful rather than beneficial to the individual. For ex ample, many inherited human diseases result from mutated genes. In addition, diseases such as skin and lung cancer can be caused by environmental agents that are known to cause DNA mutations. For these and many other reasons, understanding the molecular nature of mutations is a deeply compelling area of research.

In this chapter, we will consider the nature of mutations and their consequences on gene expression at the molecular level.

Since most mutations are harmful, organisms have developed several ways to repair damaged DNA. DNA repair systems reverse DNA damage before it results in a mutation that could potentially cause cell or organism lethality. DNA repair systems have been studied extensively in many organisms, particularly *Escherichia coli*, yeast, and mammals. A variety of systems exist that repair different types of DNA lesions. In this chapter, we will examine the ways that several of these DNA repair systems operate. First, though, we will discuss mutation and its consequences.

Consequences of Mutation

To understand why DNA mutations are beneficial or detrimental, we must appreciate how changes in DNA structure can ultimately affect DNA function. Much of our understanding of mutation has come from the study of experimental organ isms, such as bacteria, yeast, and *Drosophila*. Researchers can expose these organ isms to environmental agents that cause mutation and then study the consequences of the induced mutations. In addition, since these organisms have a short generation time, researchers can investigate the effects of mutation when they are passed from parent to offspring.

Changes in chromosome structure are referred to as *chromosome mutations*, and changes in chromosome number are called *genome mutations*. Both chromosome and genome mutations can usually be seen with the aid of a light microscope. They are important occurrences within natural populations of many eukaryotic organisms. By comparison, *single gene mutations* are relatively small changes in DNA structure that occur within a particular gene. In this chapter, we will be concerned primarily with the ways that mutations may affect the molecular and phenotypic expression of single genes. We will also consider how the timing of mutations during an organism's development has important consequences.

Gene Mutations are Molecular Changes in the DNA Sequence of a Gene

A gene mutation occurs when the sequence of the DNA within a gene is altered in a permanent way. A gene mutation can change the base sequence within a gene, or it can involve a removal or addition of one or more nucleotides. A *point mutation* is a change in a single base pair within the DNA. For example, the DNA sequence shown here has been altered by a *base substitution*:

5'-AACGCTAGATC-3'	→	5'-AACGCGAGATC-3'
3'-TTGCGATCTAG-3'		3'-TTGCGTCTAG-5'

A change of a pyrimidine to another pyrimidine (C → T) or a purine to an other purine (A → G) is called a *transition*. This type of mutation is more common than a *transversion*, in which a purine is interchanged with a pyrimidine. The example just shown is a transversion (T → G change), not a transition.

Besides base substitutions, a short sequence of DNA may be deleted from or added to the chromosomal DNA:

5'-AACGCTAGATC-3'	→	5'-AACGCTC-3'	(deletion of 4
3'-TTGCGATCTAG-3'		3'-TTGCGAG-5'	base pairs)
5'-AACGCTAGATC-3'	→	5'-AACAGTCGCTAGATC-3'	(addition of
3'-TTGCGATCTAG-3'		3'-TTGTCAGCGATCTAG-5'	4 base pairs)

As we will see next, small deletions or additions to the sequence of a gene can significantly affect its function.

Gene Mutations can Alter the Coding Sequence within a Gene

The occurrence of a mutation within the coding sequence of a structural gene can have various effects on the amino acid sequence of the polypeptide encoded by a gene. *Silent mutations* are those that do not alter the amino acid sequence of the polypeptide even though the nucleotide sequence has changed. Since the genetic code is degenerate, silent mutations can occur in the wobble base so that the type of amino acid is not changed. In contrast, *missense mutations* are base substitutions in which an amino acid change does occur. An example of a missense mutation occurs in the human disease known as sickle cell anemia. This disease involves a mutation in the β-globin gene. In the most common form of this disease, a missense mutation alters the polypeptide sequence so that the sixth amino acid is changed from a glutamic acid to valine. This single amino acid substitution alters the structure and function of the hemoglobin protein. One consequence of this alteration is that the red blood cells sickle under conditions of low oxygen.

Nonsense mutations involve a change from a normal codon to a termination codon. This causes the polypeptide to be terminated earlier than expected, producing a truncated polypeptide. Finally, *frameshift mutations* involve the addition or deletion of nucleotides in multiples of one or two. Since the codons are read in multiples of three, this shifts the reading frame so that a completely different amino acid sequence occurs downstream from the mutation.

Except for silent mutations, most new mutations are likely to produce polypeptides that have reduced rather than better function. For example, nonsense mutations will produce polypeptides that are substantially shorter and, therefore, unlikely to function properly. Likewise, frameshift mutations dramatically alter the amino acid sequence of polypeptides and are thereby likely to disrupt function. Missense mutations are less likely to alter function, since they involve a change of a single amino acid within polypeptides that typically contain hundreds of amino acids. When a missense mutation has no detectable effect on protein function, it is referred to as a *neutral mutation*. Silent mutations are also considered neutral mutations.

Mutations can occasionally produce a polypeptide that has an enhanced ability to function. While these favorable mutations are relatively rare, they may result in an organism with a greater likelihood to survive and reproduce. If this is the case, natural selection may cause such a favorable mutation to increase in frequency within a population.

Gene Mutations are also Given Names that Describe how they Affect the Wild-type Genotype and Phenotype

Thus far, we have introduced several genetic terms that describe the molecular effects of mutations. Genetic terms are also used to describe the effects of mutations relative to a wild-type genotype. In a natural population, the *wild-type* is the most common genotype. For example, the most common form of the β-globin gene is called the wild-type allele. For some genes, there may be multiple alleles that are prevalent in a population, so that there is not a single wild-type allele.

A *forward mutation* changes the wild-type genotype into some new variation. For example, in the sickle cell allele of the β-globin gene, the sixth amino acid is changed from a glutamic acid to a valine. The mutation is "forward" in an evolutionary sense, since it has changed from the prevalent genotype in the population. If a forward mutation is beneficial, it may move evolution forward; otherwise, it will probably be eliminated from the population. A *reverse mutation* has the opposite effect. For example, if the sickle cell allele mutated back to the wild-type allele (valine to glutamic acid), this would be a reverse mutation or a *reversion*.

Another way to describe a mutation is based on its influence on the wild-type phenotype. When a mutation alters the phenotypic characteristics of an organism, it is said to be a *variant*. Variants are often characterized by their differential ability to survive. As mentioned, a neutral mutation does not alter protein function, so it does not affect survival. A *deleterious mutation* will decrease the chances of survival. The extreme example of deleterious mutation is a *lethal mutation*, which results in death to the cell or organism. On the other hand, a *beneficial mutation* will enhance the survival of the organism. In some cases, an allele may be either deleterious or beneficial depending on the genotype and/or the environmental conditions. An example is the sickle cell allele, In the homozygous state, the sickle cell allele lessens the chances of survival. However, when an individual is heterozygous for the sickle cell allele and wild-type allele, this increases the chances of survival due to malarial resistance. Finally, some mutations are called *conditional mutants* because they only affect the phenotype under a defined set of conditions. Geneticists often study conditional mutants in microorganisms; a common example are temperature-sensitive (*ts*) mutants. A bacterium that has a *ts* mutation will grow normally in one temperature range (i.e., the permissive temperature) but will exhibit defective growth at a different temperature range (i.e., the non permissive temperature). For example, an *E. coli* strain carrying a *ts* mutation may be able to grow at 37°C but not at 42°C, whereas the wild-type strain can grow at either temperature.

A second mutation will sometimes affect the phenotypic expression of a first mutation. For example, a forward mutation may cause an organism to grow very slowly. A second mutation at another site in the organism's DNA may restore the normal growth rate, converting the original mutant back to the wild-type condition, Geneticists call these second site mutations *suppressors* or *suppressor mutations*. This name reflects that a suppressor mutation acts to suppress the phenotypic effects of another mutation. A suppressor mutation differs from a reversion, because it occurs at a DNA site that is distinct from that of the first mutation.

Suppressor mutations are classified according to their relative locations with regard to the mutation they suppress. When the second mutant site is within the same gene as the first mutation, it is termed an *intragenic suppressor*. Alternatively, a suppressor mutation can be in a different gene from the first mutation; this is called an *intergenic suppressor*.

There are two general types of intergenic suppressors: those that involve an ability to defy the genetic code and those that involve a mutant structural gene. A common example of the first type are suppressor tRNA genes, which have been identified in microorganisms. Suppressor tRNA genes have a change in the anti codon region that causes the tRNA to behave contrary to the genetic code. For example, nonsense suppressors are mutant tRNAs that recognize a stop codon and put an amino acid into the growing- chain. This type of mutant tRNA can suppress a nonsense mutation in another gene.

A second type of intergenic suppressor mutation are those that occur within structural genes. These suppressor mutations usually involve a change in the expression of one gene that compensates for a defective mutation affecting another gene. For example, a first mutation may cause one protein to be partially or completely defective. An intergenic suppressor mutation in a different structural gene might overcome this defect by altering the structure of a second protein so that it could take over the functional role that the first protein cannot perform. Alternatively, intergenic suppressors may involve proteins that participate in a common cellular function. When a first mutation decreases the activity of a protein, a suppressor mutation could enhance the function of the second protein involved in this common function and thereby overcome the defect in the first protein. Interestingly, intergenic suppressors sometimes involve mutations in genetic regulatory proteins such as transcription factors. When a first mutation causes a protein to be defective, a suppressor mutation may occur in a gene that encodes a transcription factor. The mutant transcription factor transcriptionally activates other genes that can compensate for the loss of function mutation in the first gene.

Gene Mutations can Occur Outside of the Coding Sequence and still Influence Gene Expression

Thus far in this chapter, we have focused our attention primarily on mutations in the coding regions of genes and their effects on gene expression. A mutation can occur within non coding sequences and thereby affect gene expression. For example, a mutation may alter the sequence within the core promoter of a gene. If the mutant promoter sequence becomes more like the consensus sequence, the mutation may increase the rate of transcription. This is called an *up-promoter mutation*. In contrast, a *down-promoter mutation* occurs when a mutation causes the promoter to become less like the consensus sequence, decreasing its affinity for regulatory factors and decreasing the transcription rate.

For example, mutations in the *lac* operator site, called *lacO*^{C} mutants, prevent the binding of the *lac* repressor protein. This causes the *lac* operon to be constitutively expressed even in the absence of lactose. Bacteria strains with *lacO*^{C} mutations are at a selective disadvantage compared with wild-type *E. coli* strains, because they waste their energy expressing the *lac* operon even when these proteins are not needed. Mutations can also occur in other noncoding regions of a gene and alter gene expression in a way that may affect phenotype. For example, mutations in eukaryotic genes can alter splice junctions and affect the order and/or number of exons that are contained within mRNA. In addition, mutations that affect the untranslated regions of mRNA (i.e., 5'— and 3'—UTRs) may affect gene expression if they alter the stability of mRNA or its ability to be translated.

DNA Sequences Known as Trinucleotide Repeats may cause Mutation

Researchers have discovered several human genetic diseases caused by an unusual form of mutation known as *trinucleotide repeat expansion* (TNRE). These diseases include fragile l syndrome (FRAXA), FRAXE mental retardation, myotonic muscular dystrophy (DM), spinal and bulbar muscular atrophy (SBMA), Huntington disease (HD), and spinocerebellar ataxia (SCA1).

There are two particularly unusual features that TNRE disorders have in common. First, the severity of the disease tends to worsen in future generations. This phenomenon is called *anticipation*. A second perplexing feature of TNRE disorders is that the severity of the disease depends on whether the disease is inherited from the mother or father. In the case of Huntington disease, TNRE is likely to occur if inheritance occurs from the father. In contrast, myotonic muscular dystrophy is more likely to get worse if it is inherited from the mother. Overall, TNRE is a newly discovered form of mutation that is receiving a lot of attention by the re search community. It poses many challenging questions in molecular genetics. TNRE also makes it particularly difficult for genetic counselors to advise couples as to the severity of these diseases if they are passed to their children.

To understand why TNRE occurs, we need to take a molecular look at the expansion process. As the name suggests, trinucleotide repeat expansion involves in creased repetition of a trinucleotide sequence. In normal individuals, certain genes and chromosomal locations contain regions where trinucleotide sequences are repeated in tandem. These sequences are transmitted normally from parent to offspring without mutation. However, in persons with TNRE disorders, the length of a trinucleotide repeat has increased above a certain critical size and becomes prone to frequent expansion. This phenomenon is depicted in the following, where the trinucleotide repeat of CAG has expanded from 11 tandem copies to 18 copies:

—CAGCAGCAGCAGCAGCAGCAGCAGCAGCAGCAG–n = 11

↓

—CAGCAGCAGCAGCAGCAGCAGCAGCAGCAGCAGCAGCAG
CAGCAGCAGCAGCAG–n = 18

The cause of TNRE is not well understood. It has been speculated that the trinucleotide repeat produces alterations in DNA structure, such as stem-loop formation, and this may lead to errors in DNA replication. However, future research will be necessary to understand the underlying mechanism that causes TNRE. Nevertheless, it is well established that TNRE within certain genes alters the expression of the gene and thereby produces the disease symptoms.

Changes in Chromosome Structure can Affect the Expression of a Gene

The mutations that we have previously considered in this chapter have been small changes in the DNA sequence of particular genes. However, a change in chromo some structure can also be associated with an alteration in the expression of a specific gene. Quite commonly, an inversion or translocation has no obvious phenotypic consequence. However, Alfred Sturtevant recognized as early as 1925 that chromosomal re arrangements in *Drosophila melanogaster* can influence phenotypic expression (namely, eye morphology). In some cases, a chromosomal rearrangement may affect a gene because the chromo-

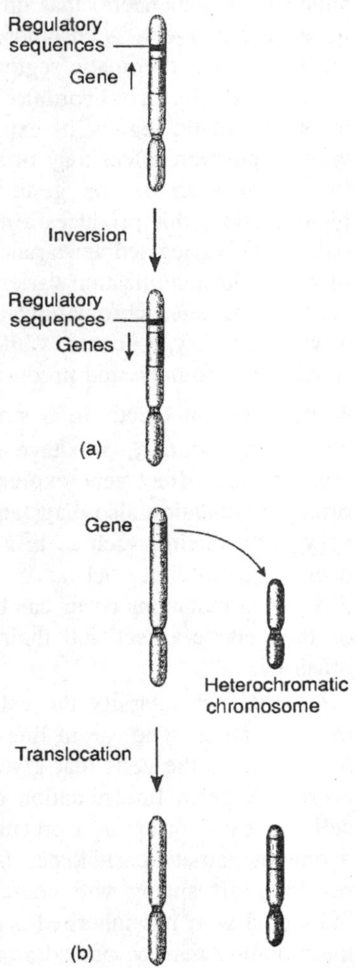

Fig. 8.1. Causes of position effects. (a) A chromosomal inversion, (b) a translation.

somal break-point actually occurs within the gene it self. In other cases, a gene maybe left intact, but its expression may be altered when it is moved to a new location. When this occurs, the change in gene location is said to have a *position effect*.

There are two common reasons for position effects. One possibility is that a gene may be moved next to regulatory sequences (i.e., silencers or enhancers) that influence the expression of the relocated gene Alternatively, a chromosomal rearrangement may reposition a gene from a euchromatic region to a chromosome that is very highly condensed (heterochromatic). When the gene is moved to a heterochromatic region, its expression maybe turned off. This second type of position effect may produce a variegated phenotype in which the expression of the gene is variable. For genes that affect pigmentation, this produces a mottled appearance rather than an even color. The variegated appearance of the eye occurs because the degree of heterochromatinization varies across different regions of the eye. In cells where heterochromatinization has turned off the eye color gene, a white phenotype occurs, while other cells allow this same region to remain euchromatic and produce a red phenotype.

Mutations can Occur in Germ line or Somatic Cells

In this section, we have considered many different ways that mutation can affect gene expression. For multicellular organisms, the timing of mutation also plays an important role .A mutation can occur very early in life, such as in a gamete or a fertilized egg, or it may occur later in life, such as in embryonic or adult stages. The exact time when mutations occur can be important with regard to the severity of the genetic effect and their ability to be passed from parent to offspring.

Geneticists classify the cells of sexually reproducing organisms into two types: The germ line and the somatic cells. The term *germ line* refers to the cells that give rise to the gametes such as eggs and sperm. A germ line mutation can occur directly in a sperm or egg cell, or it can occur in a precursor cell that produces the gametes. If a mutant gamete participates in fertilization, all of the cells of the resulting off spring will contain the mutation. Likewise, when an individual who has inherited a germ line mutation produces gametes, the mutation may be passed along to future generations of offspring.

The *somatic* cells comprise all cells of the body excluding the germ line cells. Examples include muscle cells, nerve cells, and so forth. Mutations can also occur within somatic cells at early or late stages of development. In this example, a mutation has occurred within a single embryonic cell. As the embryo grows, this single cell will be the precursor for many cells of the adult organism. Therefore, in the adult, a patch of tissue will contain the mutation. The size of the patch will depend on the timing of the mutation. In general, the earlier the mutation occurs during development, the larger the patch. An individual who has somatic regions that are genotypically different from each other is called a *genetic mosaic*.

In this case, the person has a patch of gray hair while the rest of the hair is pigmented. Presumably, this individual initially had a single mutation occur in a embryonic cell that ultimately gave rise to a patch of scalp that produced the gray hair. Although a patch of gray hair is not a particularly harmful phenotypic effect, mutations during early stages of life can be quite harmful, especially if they disrupt essential developmental processes. Therefore, even though it is prudent to avoid environmental agents that cause mutations during all stages of life, the possibility of somatic mutations is a rather compelling reason to avoid them during the very early stages of life such as fetal development, infancy, and early childhood.

Occurrence and Causes of Mutation

As we have seen, mutations can have a wide variety of effects on the phenotypic expression of genes. For this reason, geneticists have spent a great deal of effort identifying the causes of mutations. This has been a truly challenging task, since various agents can alter the structure of DNA and thereby cause mutation. Geneticists categorize the cause of mutation in one of two ways. *Spontaneous mutations* are changes in DNA structure that result from abnormalities in biological processes, whereas *induced mutations* are caused by environmental agents.

Many causes of spontaneous mutations are examined in other chapters throughout this text. Abnormalities in crossing over can produce chromosome mutations such as deletions, duplications, translocations, and inversions. Aberrant segregation of chromosomes during meiosis can cause genome mutations (i.e., changes in chromosome number). In We discovered that DNA polymerase can make a mistake during DNA replication by putting the wrong base in a newly synthesized daughter strand. Errors in DNA replication are usually in frequent except in certain viruses, such as HIV, that have relatively high rates of spontaneous mutations. In addition, normal metabolic processes may produce chemicals within the cell that can react directly with the DNA and alter its structure. Also, as will be described, transposable genetic elements can alter gene sequences by inserting themselves into genes. Overall, a distinguishing feature of spontaneous mutations is that their underlying cause originates within the cell.

By comparison, the cause of induced mutations originates outside of the cell. Induced mutations are produced by environmental agents that enter the cell and then alter the structure of DNA. Agents that are known to alter the structure of DNA are called *mutagens*. Mutagens can be chemical substances or physical agents that ultimately lead to changes in DNA structure.

In this section, we will begin by examining the random nature of spontaneous mutations and some general features of the mutation rate. We will then explore several mechanisms by which mutagens can alter the structure of DNA. Laboratory tests that can identify potential mutagens will then be described.

Spontaneous Mutations are Random Events

For many centuries, biologists have wondered whether genotypic changes occur purposefully as a result of environmental conditions or whether they are spontaneous events that may occur randomly in any gene of any individual. The question of whether such mutations are spontaneous occurrences or causally related to environmental conditions has an interesting history. In the 19th century, Jean Baptiste Lamarck proposed that physiological events (e.g., use and disuse) determine whether traits are passed along to offspring. For example, his theory suggested that an individual who practiced and became adept at a physical activity; such as the long jump, would pass that quality on to their offspring. Alternatively, Charles Darwin proposed that genetic variation occurs as a matter of chance and that natural selection results in the differential survival of organisms that are better adapted to their environments. According to this view, those individuals who happen to contain beneficial mutations will be more likely to survive and pass these genes to their offspring. These opposing theories of the 19th century were tested in bacterial studies in the 1940s and 1950s. Two of these studies are described here.

Salvadore Luria and Max Delbruck, while working at Indiana and Vanderbilt Universities, were interested in the ability of bacteria to become resistant to infection by a bacteriophage called T1. When a population of *E. coli* cells is exposed to T1, a small percentage of bacteria become resistant to T1 infection and pass this trait to their progeny. Luria and Delbruck were interested in whether such resistance, called *Ton*^{R} (T one Resistance), is due to the occurrence of spontaneous mutations or whether it is a physiological adaptation that occurs at a low rate within the bacterial population.

According to the physiological adaptation theory; the rate of adaptation should be a relatively constant value and would depend on the exposure to the bacteriophage. Therefore, when comparing different populations of bacteria, the number of *Ton*^{R} bacteria should be an essentially constant proportion of the total population. In contrast, the spontaneous mutation theory depends on the timing of mutation. If a *Ton*^{R} mutation occurs early within the proliferation of a bacterial population, many *Ton*^{R} bacteria will be found within that population. However, if it occurs much later in population growth, then fewer *Ton*^{R} bacteria will be observed. In general, a spontaneous mutation theory predicts a much greater fluctuation in the number of *Ton*^{R} bacteria among different populations. This test, therefore, has become known as the *fluctuation test*.

To distinguish between the physiological adaptation and spontaneous mutation theories, Luria and Delbruck inoculated 20 individual tubes and one large flask with *E. coli* cells and grew them in the absence of T1 phage. The flask was grown to produce a very large population of cells, while each individual culture was grown to a smaller population of approximately 20 million cells. They then plated the individual cultures onto media containing T1 phage. Likewise, 10 sub-samples, each consisting of 20 million bacteria, were removed from the large flask and plated onto media with T1 phage.

Within the smaller individual cultures, a great fluctuation was observed in the number of *Ton*^{R} mutants. These results are consistent with a spontaneous mutation theory in which the timing of a mutation during the growth of a culture greatly affects the number of mutant cells. For example, in tube #14 there were many *Ton*^{R} bacteria. Luria and Delbruck reasoned that a mutation occurred randomly in one bacterium at an early stage of the population growth, before the bacteria were exposed to T1 on plates. This mutant bacterium then divided to produce many daughter cells that inherited the *Ton*^{R} trait. In other tubes, such as #1 and #3, this spontaneous mutation did not occur, and so none of the bacteria had a *Ton*^{R} phenotype. By comparison, the cells plated from the large flask tended toward a relatively constant and intermediate number of *Ton*^{R} bacteria. Since the large growth flask had so many cells, several independent *Ton*^{R} mutations were likely to have occurred during different stages of its growth. In a single flask, however, these independent events would be mixed together to give an average value of *Ton*^{R} cells.

Randomly Occurring Mutations can give an Organism a Survival Advantage

Joshua Lederberg and Ester Lederberg were also interested in the relationship between mutation and the environmental conditions that select for mutation. At the time of their studies, some scientists still held the belief that selective conditions could promote the formation of specific mutations allowing the organism to survive. Similar to the Lamarck theory of adaptive mutations, these scientists thought that if bacteria were exposed to an antibiotic, for example, the presence of the antibiotic would actually cause gene mutations that confer antibiotic resistance. This theory, which was similar to the adaptation theory, was known as the *directed mutation theory*. In contrast, *the random mutation theory*, which was consistent with a Darwinian viewpoint, proposed that mutations occur at random. According to this theory, environmental factors that affect survival simply select for the survival of those individuals that happen to possess a beneficial mutation.

An experimental approach to distinguish between these two possibilities was developed by the Lederbergs in the 1950s while working at the University of Wisconsin in Madison. They used a technique known as *replica plating*. They plated a large number of bacteria onto a master plate that did not contain any selective agent (namely, T1).

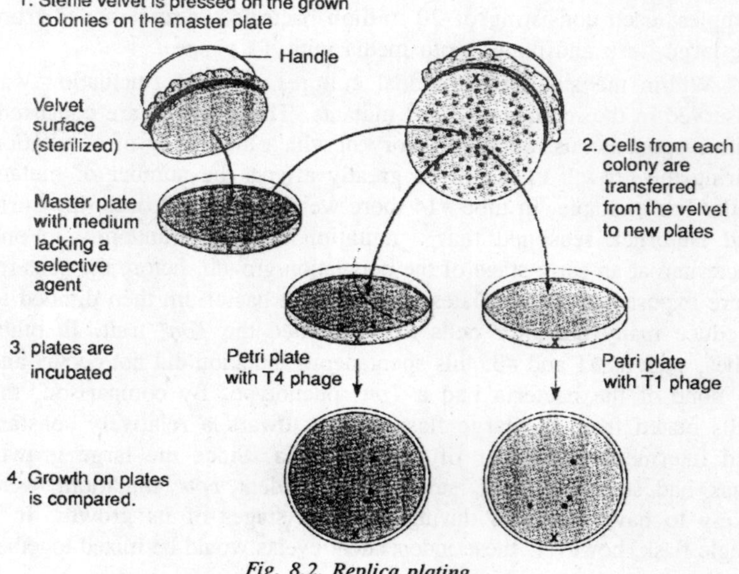

Fig. 8.2. Replica plating.

A sterile piece of velvet cloth was lightly touched to this plate in order to pick up a few bacterial cells from each colony. This replica was then transferred to secondary plates that contained an agent that selected for the growth of bacterial cells with a particular genotype.

The secondary plates contained T1 bacteriophages. On these plates, only those mutant cells that are *Ton*^{R} could grow. On the secondary plates, a few colonies were observed. Furthermore, they occupied the same location on each plate. These results indicated that the mutations conferring *Ton*^{R} occurred randomly while the cells were growing on the (nonselective) master plate; the presence of the bacteriophage in the secondary plates simply selected for the growth of previously occurring *Ton*^{R} mutants. These results supported the random mutation theory. In contrast, the directed mutation theory would have predicted that *Ton*^{R} bacterial mutants would occur after the cells were transferred to the secondary plates. If that had been the case, they would not be expected to arise in identical locations on different secondary plates.

The Mutation Rate is a Measure of New Mutations per Generation; the Mutation Frequency is the Relative Occurrence of a Mutation within a Population

Since mutations occur spontaneously among populations of living organisms, geneticists have been greatly interested in learning how prevalent they are. The term *mutation rate* is the likelihood that a gene will be altered by a new mutation. It is commonly expressed as the number of new mutations in a given gene per generation. In unicellular organisms such as bacteria, the spontaneous mutation rate for a particular gene is in the range of 1 in 100,000 to 1 in 1,000,000,000, or 10^{-5} to 10^{-9} per cell generation. In more complex eukaryotes, the rate of germ line mutations can be expressed as the number of new mutations per generation. These numbers tell us that it is very unlikely that a particular gene will mutate due to natural causes. However, the mutation rate is not a constant number. The presence of mutagens within the environment can increase the rate of induced mutations to a much higher value than the spontaneous mutation rate. In addition, mutation rates vary substantially from species to species and even within different strains of the same species. One explanation for this variation is the many different causes of mutations.

As we have learned more about mutation rate, it has been necessary to modify the view that mutations are a totally random process. Within the same individual, different genes can vary widely with regard to their individual mutation rates. Some genes mutate at a much higher rate than other genes. This is because some genes are larger than others (providing a greater chance for mutation) and their lo cations within the chromosome may be more susceptible to mutation. Even within a single gene, there are usually *hot spots*, which are select regions of a gene that are more likely to mutate than are other regions.

Before we end our discussion of mutation rate, it is helpful to distinguish the rate of new mutation from the concept of *mutation frequency*. The mutation frequency for a gene is the number of mutant genes divided by the total number of genes within the population. If 1,000,000 bacteria were plated and 10 were found to be mutant, the mutation frequency would be 1 in 100,000 or 10^{-5}. As described earlier in the chapter, Luria and Delbruck showed that among the bacteria in the 20 tubes, the timing of mutations influenced the mutation frequency within any particular tube. Some tubes had a high frequency of mutation, while others did not. The mutation frequency is an important genetic concept, particularly in the field of population genetics. The mutation or allelic frequency depends on a variety of factors, including the mutation rate and the forces of natural selection.

X-rays were the first environmental agent shown to cause induced mutations, by Muller in 1927

Geneticists have used allelic variation (e.g., red versus white eyes) as an important tool to explore the mechanisms of inheritance. In natural populations, allelic variation can occur as a result of spontaneous mutation. As we have just discussed, though, the rate of spontaneous mutations for any given gene tends to be very low, in the range of one in a million. From an experimental viewpoint, this low rate of spontaneous mutation presents an obstacle to the study of allelic variation among individuals. For example, it is not an easy task to observe mil lions of fruit flies in search of rare spontaneous mutations that cause a phenotypic variation. Thomas Hunt Morgan spent two years before he obtained his first white-eye mutation. In response to this obstacle, many geneticists have sought to identify environmental agents that promote mutation and thereby make it easier to study mutation in experimental organisms.

In 1927, long before the structure of DNA was known, Hermann Muller at the University of Texas devised an approach that demonstrated that X-rays can cause mutation. This work was carried out using *Drosophila melanogaster* as an experimental organism. Muller reasoned that a mutagenic agent might cause some genes to become defective. His experimental approach focused on the ability of a mutagen to cause defects in X-linked genes that result in a recessive, lethal phenotype. In particular, the goal of Muller's experiment was to determine whether X-rays in crease the production of X-linked recessive mutations. To determine if X-rays increase the rate of recessive X-linked lethal mutations, Muller wanted to have an easy way to detect the occurrence of such mutations. He cleverly realized that he had available a laboratory strain of fruit flies that could make this possible. In particular, he conducted his crosses in such a way that a female fly that inherited a new mutation causing a recessive X-linked lethal allele would not be able to produce any male offspring. This made it very easy for him to detect lethal mutation; he only had to count the number of female flies that could not produce sons. In flies that had not been exposed to X-rays, nearly all females were able to produce sons. In contrast, if X-rays were acting as a mutagenic agent, Muller hypothesized that exposure to X-rays would increase the numbers of females unable to produce sons.

To understand Muller's crosses, we need to take a closer look at a peculiar version of one of the X-chromosomes in a strain of flies that he used in his crosses. This X-chromosome, designated ClB, had three important genetic alterations:

1. It contained a large inversion that prevented it from Crossing over with the other X-chromosome in female files. The letter C is a reminder that this region of the chromosome cannot cross over.
2. It carried a lethal recessive X-linked gene. If male (XY) inherit this chromosome, they will die.
3. It carried a dominant mutation that causes the eyes of the fly to have a **B**ar shape.

Note: C and B are uppercase because they are inherited in a dominant manner, while l is lowercase because it is a recessive allele.

A female fly that has one copy of this X-chromosome would have bar-shaped eyes. Even though this X-chromosome has a lethal allele, a female fly can survive if the corresponding gene on the other X-chromosome is a normal allele. In Muller's experiments, the goal was to determine if the normal X-chromosome (not the ClB chromosome) had obtained a lethal mutation in any gene except for the same lethal gene on the ClB chromosome. If a recessive lethal mutation occurred on the normal X-chromosome, this female could survive because it would be heterozygous for recessive lethal mutations in two different genes. However, since each X-chromosome would have a lethal mutation, this female would not be able to produce any living sons.

Hypothesis

The exposure of flies to X-rays will increase the rate of mutation.

Starting material: The female flies contain one normal X-chromosome and on ClB X-chromosome. The male flies contain a normal X-chromosome.

1. Expose male flies to X-rays. Also, have a control group that is not exposed to X-rays.
2. Mate the male flies to female flies carrying one normal X-chromosome and one ClB X-chromosome.
3. Save about 1000 daughters with bar eyes. Note: These females contain a ClB X-chromosome from their mothers and an X-chromosome from their fathers that may (or may not) have a recessive lethal mutation.
4. Mate each bar-eyed daughter with normal (non irradiated) males. Note: This is done in (1000) individual tubes.
5. Count the number of crosses that do not contain any male offspring. These crosses indicate that the bar-eyed female parent contained an X-linked recessive mutation on the non-ClB X-chromosome.

Interpreting the data

In the absence of X-ray treatment, only one cross in approximately 1000 was unable to produce male offspring. This means that the spontaneous rate for any X-linked lethal mutation was relatively low. By comparison, X-ray treatment (of the fathers) of the ClB females produced 91 crosses without male offspring. Since these females inherited their non-ClB chromosome from irradiated fathers, these results indicate X-rays greatly increase the rate of X-linked recessive lethal mutations. This conclusion has been confirmed in many subsequent studies, which have shown that the increase in mutation rate is correlated with the amount of exposure to X-rays.

Mutagens Alter DNA Structure in Different Ways

Since this pioneering study of Muller, researchers have found that an enormous array of agents can act as mutagens to permanently alter the structure of DNA. We often hear in the news media that we should avoid these agents in our foods and living environment. We even use products such as sunscreens that help us avoid the mutagenic effects of ultraviolet (UV) rays. The public is concerned about mutagens for two important reasons. First, mutagenic agents are often involved in the development of human cancers. In addition, since most new mutations are deleterious, people want to avoid mutagens to prevent gene mutations that may have harmful effects in their offspring. Mutagenic agents are usually classified as *chemical* or *physical* mutagens. In other cases, chemicals that are not mutagenic can be altered to a mutagenically active form after they have been ingested into the body. Cellular enzymes such as oxidases have been shown to activate some mutagens. Certain foods contain chemicals that act as antioxidants. Many scientists are investigating whether antioxidants can counteract the effects of mutagens and thereby lower the cancer rate.

Mutagens can alter the structure of DNA in various ways. Some mutagens act by covalently modifying the structure of nucleotides. For example, *nitrous acid* replaces amino groups with keto groups ($—NH_2$ to = 0). This can change cytosine to uracil, and adenine to hypoxanthine. When this mutated DNA replicates, the modified bases do not pair with the appropriate nucleotides in the newly made strand. Instead, uradil pairs with adenine, and hypoxanthine pairs with cytosine. Other chemical mutagens can also disrupt the appropriate pairing between nucleotides by alkylating bases within the DNA. During alkylation, methyl or ethyl groups are covalently attached to the bases. Examples of alkylating agents include *nitrogen mustards* and *ethyl methanesulfonate* (EMS).

Some mutagens exert their effects by directly interfering with the DNA replication process. For example, *acridine dyes*, such as *proflavin*, contain flat planar structures that insert themselves into the double helix, thereby distorting the helical structure. When DNA containing these mutagens is replicated, single nucleotide additions and/or deletions can be incorporated into the newly made daughter strands.

Compounds such as *5-bromouracil* (5BU) and *2-aminopurine* are nucleotide base analogues that become incorporated into daughter strands during DNA replication. 5-Bromouracil is a thymine analogue that can be incorporated into DNA instead of thymine. Like thymine, 5BU can base pair with adenine. However, at a relatively high rate, it can also base pair with guanine. When this occurs during DNA replication, 5BU causes a mutation in which an A—T base pair is changed to a G—5BU base pair. This is a transition, since the adenine has been changed to a guanine, both of which are purines. During the next round of DNA replication, the template strand containing the guanosine base will create a G—C base pair. In this way, 5-bromouradil can promote a change of an A—T base pair into a G—C base pair.

DNA molecules are also sensitive to physical agents such as radiation. In particular, radiation of short wavelength and high energy is known to alter DNA structure. Ionizing radiation includes X-rays and gamma rays. This type of radiation can penetrate deeply into biological materials, where it creates chemically reactive molecules known as *free radicals*. These molecules can alter the structure of DNA in a variety of ways. Exposure to high doses of ionizing radiation can cause base deletions, single nicks in DNA strands, cross-linking, and even chromosomal breaks. Nonionizing radiation, such as UV light, contains less energy; and so it only penetrates the surface of material such as the skin. Nevertheless, UV light is known to cause DNA mutation. For example, as shown in Figure, UV light causes the formation of cross-linked *thymine dimers*. A thymine dimer within a DNA strand may cause a mutation when that DNA strand is replicated.

Testing Methods can Determine if an Agent is a Mutagen

To determine if an agent is mutagenic, researchers utilize testing methods that can monitor whether or not an agent increases the rate of mutation. Many different kinds of tests have been used to evaluate mutagenicity. One commonly used test is the *Ames test*, which was developed by Bruce Ames at UC Berkeley in the 1970s. This test uses strains of a bacterium, *Salmonella typhimurium*, that cannot synthesize the amino acid histidine. These strains contain a deleterious mutation within a gene that encodes an enzyme required for histidine biosynthesis. Therefore, the bacteria cannot grow on petriplates unless histidine has been added to the growth medium. However, a second mutation may occur that restores the ability to synthesize histidine. In other words, a second mutation can cause a reversion back to the wild-type condition. The Ames test monitors the rate at which this second mutation occurs and thereby indicates whether an agent increases the mutation rate above the spontaneous rate.

The suspected mutagen is mixed with a rat liver extract and a bacterial strain that can't synthesize histidine. A mutagen may require activation by cellular enzymes; the rat liver extract provides a mixture of enzymes that may activate a mutagen. This step improves the ability of the test to identify agents that may cause mutation in mammals. After the incubation period, a large number of bacteria are then plated on a minimal growth medium that does not contain histidine. The *Salmonella* strain is not expected to grow on these plates. However, if a mutation has occurred that allows a bacterium to synthesize histidine, it can grow on these plates to form a visible bacterial colony. To estimate the mutation rate, the colonies that grow on the minimal media are counted and compared with the total number of bacterial

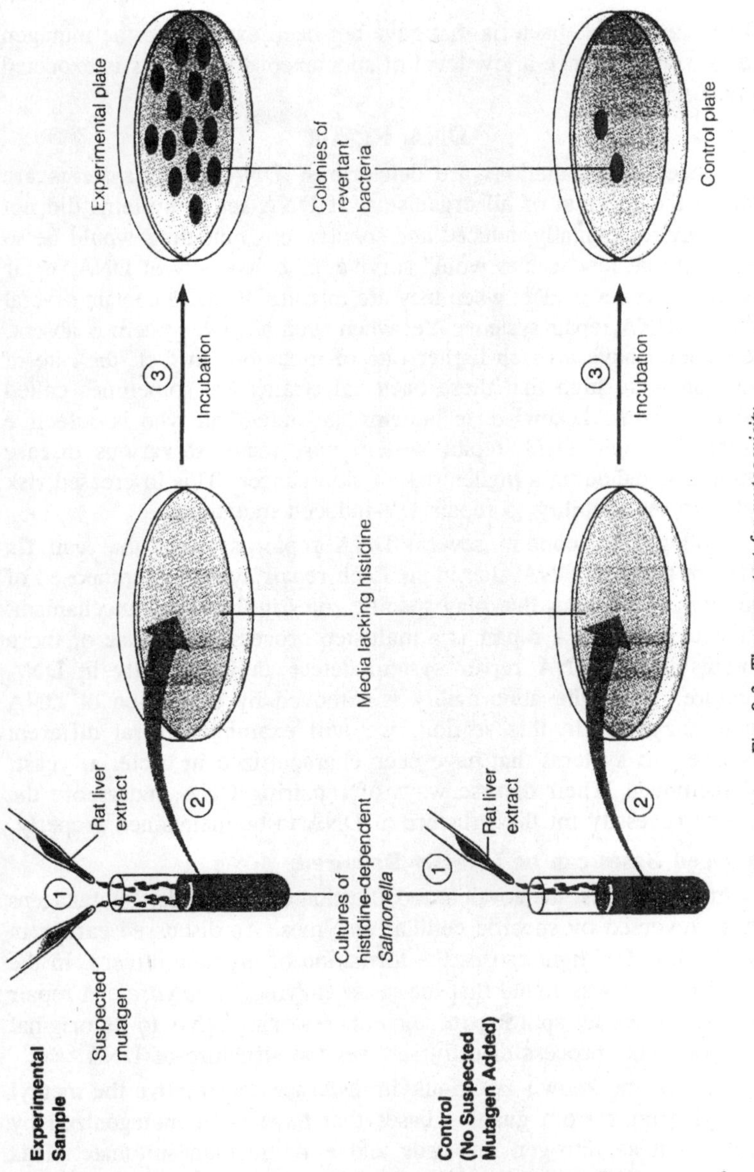

Fig. 8.3. The ames test for mutagenicity.

cells that were originally streaked on the plate. For example, if 10,000,000 bacteria were plated and 10 growing colonies were observed, the rate of mutation is 10 out of 10,000,000; this equals 1 in 10^6, or

10^{-6} As a control, bacteria that have not been exposed to the mutagen are also tested, since a low level of spontaneous mutations is expected to occur.

DNA Repair

Since most mutations are deleterious, DNA repair systems are vital to the survival of all organisms. If DNA repair systems did not exist, environmentally induced and spontaneous mutations would be so prevalent that few species would survive. The necessity of DNA repair systems becomes evident when they are missing. Bacteria contain several different DNA repair systems. Yet when even a single system is absent, the bacteria have a much higher rate of mutation. In fact, the rate of mutation is so high that these bacterial strains are sometimes called mutator strains. Likewise, in humans, an individual who is defective in only a single DNA repair system may manifest various disease symptoms, including a higher risk of skin cancer. This in creased risk is due to the inability to repair UV-induced mutations.

Living cells contain several DNA repair systems that can fix different types of DNA alterations Each repair system is composed of one or more proteins that play specific roles in the repair mechanism. In most cases, DNA repair is a multistep process. First, one or more proteins in the DNA repair system detect an irregularity in DNA structure. Next, the abnormality is removed by the action of DNA repair enzymes. In this section, we will examine several different DNA re pair systems that have been characterized in bacteria, yeast, and mammals. Their diverse ways of repairing DNA underscore the extreme necessity for the structure of DNA to be maintained properly.

Damaged Bases can be Directly Repaired

In a few cases, the covalent modification of nucleotides by mutagens can be reversed by specific cellular enzymes. As discussed earlier in this chapter, UV light causes the formation of thymine dimers. In the early 1960s, it was found that the yeast enzyme *photolyase* can repair thymine dimers by splitting the dimers, restoring DNA to its original condition. This process directly restores the structure of DNA.

An enzyme known as O6-alkyltransferase can remove the methyl or ethyl groups from guanine bases that have been mutagenized by agents such as nitrogen mustards and ethyl methanesulfonate. This enzyme is called a transferase because it transfers the methyl or ethyl group from the base to a cysteine side chain within the alkyltransferase protein. Surprisingly, this permanently inactivates alkyl transferase—it can only be used once!

Table. 8.1. Common types of DNA repair system

System	*Description*
Direct repair	An enzyme recognizes an incorrect alteration in DNA structure and directly converts it back to a correct structure.
Excision repair	An abnormal nucleotide or base is first recognized and removed from the DNA. A segment of DNA in this region is excised, and then the complementary DNA strand is used as a template to synthesize a normal DNA strand.
Mismatch repair	Similar to excision repair except that the DNA defect is a base pair mismatch in the DNA, not an abnormal nucleotide. The mismatch is recognized, and a segment of DNA in this region is removed. The parental strand is used to synthesize a normal daughter strand of DNA.
Recombinational repair	Occurs when a DNA mutation causes a gap in synthesis during DNA replication. The gap is exchanged between the abnormal DNA and the corresponding region in the normal replicated double helix. After this occurs, it is possible to fill in the gap using the complementary DNA strand in the normal DNA strand.

Excision Repair Systems Remove Damaged Nucleotides or Bases from DNA

An important general process for DNA repair is *nucleotide excision repair* (NER). This type of system can repair many different types of DNA damage, including UV- induced damage (namely, thymine dimers), chemically modified bases, missing bases, and certain types of cross-links. In NER, several nucleotides m the damaged strand are removed from the DNA, and the intact strand is used as a template for resynthesis of a normal complementary strand. NER is found in both eukaryotes and prokaryotes, although its molecular mechanism is better understood in prokaryotic species.

In *E. coli*, the NER system requires four key proteins, designated UvrA, UvrB, UvrC, and UvrD, plus the help of DNA polymerase and DNA ligase. UvrA, B, C, and D recognize and remove a short segment of a damaged DNA strand. They are named Uvr, because they are involved in Ultraviolet light repair of pyrimidine dimers, although the UvrA—D proteins are also important in repairing chemically damaged DNA.

A protein trimer consisting of two UvrA molecules and one UvrB molecule tracks along the DNA in search of damaged DNA. Such DNA will have a distorted double helix, which is sensed by the UvrA—B complex. When a damaged segment is identified, the two UvrA proteins are released and UvrC binds to the site. Together, UvrB and UvrC make incisions in the damaged strand on both sides of the damaged site. UvrB probably makes the 3' cut and UvrC the 5' cut. After this incision process. UvrC is released and UvrD binds. UvrD is a helicase that unravels the DNA so that the short damaged segment can be released or excised from the DNA. Following the excision of the damaged DNA, DNA polymerase fills in the gap using the undamaged strand as a template. Finally, DNA ligase makes the final covalent connection between the newly made DNA and the original DNA strand.

A second type of excision repair system involves the function of an enzyme known as *DNA-N-glycosylase*. This enzyme can recognize an abnormal base and cleave the bond between it and the sugar in the DNA backbone. Depending on the organism, this repair system can eliminate abnormal bases such as uracil, N- and 3-methyladenine, 7-methylguanine, and pyrimidine dimers.

In this example, the DNA contains a uradil residue in its sequence. This could have happened by the action of a chemical mutagen (e.g., nitrous acid) that deaminates cytosine to produce uracil. N-Glycosylase recognizes a uracil within the DNA and cleaves the bond between the sugar and base. This releases the uracil base and leaves behind an apyrimidinic (AP) nucleotide. This abnormal nucleotide is recognized by a second enzyme, AP-endonuclease, which makes a cut on the 5' side. DNA polymerase (which has a 5' to 3' exonuclease activity) removes the abnormal region and, at the same time, replaces it with normal nucleotides. This process is called *nick translation* (although nick replication would be a more accurate term). Finally, DNA ligase closes the nick.

Several human diseases have been shown to involve inherited defects in genes involved in nucleotide excision repair. These include xeroderma pigmentosum (XP), Cockayne's syndrome (CS), and PIBIDS.

(PIBIDS is an acronym for a syndrome with symptoms that include photosensitivity, ichthyosis [a skin abnormality], brittle hair, impaired intelligence, decreased fertility, and short stature.) A common characteristic in all three syndromes is an increased sensitivity to sunlight because of an inability to repair UV-induced lesions. These individuals have pigmentation abnormalities, many premalignant lesions, and a high predisposition to skin cancer. They may also develop early degeneration of the nervous system.

Genetic analyses of patients with XP, CS, and PIBIDS have revealed that these syndromes result from defects in a variety of different genes that encode NER proteins. For example, xeroderma pigmentosum can be caused by defects in seven different NER genes. In all cases, individuals have a defective nucleotide excision repair pathway. In recent years, several human NER genes have been successfully cloned and sequenced. Although more research is needed to understand completely the mechanisms of DNA repair, the identification of NER genes has helped unravel the complexities of NER pathways in human cells.

Mismatch Repair Systems Recognize and Correct a Base Pair Mismatch

Thus far, we have considered several DNA repair systems that recognize abnormal nucleotide structures within DNA, including thymine dimers, all bases, and the presence of uracil in the DNA. Another type of abnormality that should not occur in DNA is a *base mismatch*. The structure of the DNA double helix obeys the A—T/G—C rule of base pairing. During the normal course of DNA replication, however, an incorrect nucleotide may be added to the growing strand by mistake. This creates a mismatch be a nucleotide in the parental and newly made strand. There are various DNA repair mechanisms that recognize and remove this mismatch. For example, DNA polymerase has a 3' to 5' proofreading ability that can detect mismatches and remove them. However, if this proofreading ability fails, cells contain several additional DNA repair systems that can detect base mismatches and fix them.

An interesting DNA repair system is the *methyl-directed mismatch repair* system that has been studied extensively in *E. coli*. This system involves the participation of several proteins that detect the mismatch and specifically remove the segment from the newly made daughter strand. Keep in mind that the newly made daughter strand contains the incorrect base, while the parental strand is normal. Therefore, an important aspect of methyl-directed mismatch repair is that it repairs specifically the newly made strand rather than the original template strand.

Three proteins, designated MutL, MutH, and MutS, detect the mismatch and direct the removal of the mismatched base from the newly made strand. These proteins are named Mut because their absence leads to a much higher mutation rate than that in normal strains of *E. coli*. A key characteristic of the MutH protein is that it can distinguish between the parental DNA strand and the daughter strand. It can do this because the MutH protein recognizes a methylated DNA sequence but does not recognize nonmethylated DNA sequences. Immediately after DNA replication, the parental strand is methylated, but it takes a short time before a newly made daughter strand is methylated. Therefore, the MutH protein specifically recognizes the parental DNA strand rather than the daughter strand.

The role of MutS is to locate mismatches. MutS forms a complex with MutL. When the MutS—MutL complex has located a DNA mismatch, it interacts with MutH by a looping mechanism. This stimulates MutH to make a cut in the non methylated DNA strand. After the strand is cut, an exonuclease digests the nonmethylated DNA strand in the direction of the mismatch and proceeds just beyond the mismatch site. This leaves a gap in the daughter strand, which is re paired by DNA polymerase and DNA ligase. The net result is that the mismatch has been corrected by removing the incorrect region in the daughter strand and then resynthesizing the correct sequence using the parental DNA as a template.

Damaged DNA can be Repaired by Recombination

Certain types of DNA damage can halt the progression of DNA replication. For example, a thymine dimer inhibits DNA replication, because it prevents base pairing in a newly made daughter strand. For this reason, a gap will be created in the region that cannot be replicated. DNA gaps are particularly harmful, since they may cause chromosomal breaks and further mutations. To avoid these consequences, the unreplicated gap may be repaired by genetic recombination.

DNA replication produces a genetically identical pair of double-stranded DNA molecules. Recombinational repair occurs while the two DNA copies are being made. To understand this process, we must pay close attention to the relationship between newly made DNA copies during replication. The strands in the two double helices are labeled A, B, C, and D. Since DNA replication makes identical copies of genetic material, strands A and C have the same DNA sequence. Strands B and D also have the same sequence and are complementary to A and C. A thymine dimer had previously occurred in the parental strand (which is now labeled strand D). The thymines in a thymine dimer cannot hydrogen bond properly to incoming nucleotides during DNA replication. Therefore, a thymine dimer disrupts the DNA replication process by creating a gap in the newly made strand after the replication fork has passed. Recombinational repair fixes this gap. This occurs in a two-step process. First, the gap is replaced with the same region in the original DNA strand: a short segment of strand A replaces the corresponding segment of strand C. This removes the gap in strand C, but it creates a gap in strand A. The next step is to fill in the gap in strand A, using the normal strand B as a template. The net result is that neither of the replicated DNA double helices contain gaps, although the thymine dimer is still present in one of the double helices.

Certain bacteria, such as *E. coli*, also have an alternative mechanism to fill in gaps that occur during the replication of damaged DNA. This mechanism, known as the *SOS response*, occurs under extreme conditions that promote DNA damage. Causes of the SOS response include high doses of UV light and other types of mutagens. When the SOS response occurs, the gaps in replicating DNA are filled in by a mechanism that does not involve recombination. When the gaps are filled in during the SOS response, a high error rate of mutation occurs. Even though this is a harmful result, it nevertheless allows the bacteria to survive under conditions of extreme environmental stress. Furthermore, the high rate of mutation may provide genetic variability within the bacterial population, so that certain cells may become resistant to the harsh conditions.

Actively Transcribing DNA is Repaired more Efficiently than Non-transcribed DNA

Before we end our discussion of DNA repair systems, it is interesting to mention that not all DNA is repaired at the same rate. In the 1980s, Philip Hanawalt and colleagues at Stanford conducted experiments showing that actively transcribing genes in eukaryotes and prokaryotes are more efficiently repaired following radiation damage than is non-transcribed DNA. The targeting of DNA repair enzymes to actively transcribing genes may have several biological advantages. First, active genes are loosely packed and may be more susceptible to DNA damage. Likewise, the process of transcription itself may cause

DNA damage. In addition, DNA regions that contain actively transcribing genes are more likely to be important for survival than non-transcribed regions. This may be particularly true in terminally differentiated cells, which no longer divide. In non-dividing cells, gene transcription (rather than DNA replication) is of utmost importance.

In eukaryotes, the mechanism that couples DNA repair and transcription is not completely understood. Certain general transcription factors, such as TFII-H, play a role in both transcription and excision repair. In *E. coli*, a protein known as *transcription—repair coupling factor* (TRCF) is responsible for targeting the excision repair system to actively transcribing genes having damaged DNA. In this scenario, RNA polymerase is actively transcribing a gene until it encounters a DNA lesion and be comes stalled. TRCF binds to the stalled RNA polymerase and displaces it from the damaged site. TRCF recruits the excision repair system to the damaged site by recognizing the UvrA—UvrB complex. After the UvrA—UvrB complex has bound to this site, TRCF is released from the damaged site. At this stage, the UvrA—B complex can begin excision repair.

Concluding Remark

Mutations are heritable changes the genetic material. They can result from small changes in the genetic material (*single gene mutations*), rearrangements in chromosome structure (*chromosome mutations*), or changes in chromosome number (*genome mutations*). Mutations can have various consequences on gene expression. If a mutation occurs within the coding sequence, it may alter the amino acid sequence of the encoded polypeptide. *Missense mutations* cause single amino acid substitutions, *nonsense mutations* result in truncated polypeptides, and *frame shift mutations* produce changes in the amino acid reading frame. In general, these types of mutations are more likely to be detrimental rather than beneficial to protein structure and function. Mutations can also occur outside of the coding sequence and significantly affect gene expression. For example, mutations within promoters, response elements, splicing sequences, and untranslated sequences can have major consequences on the level of gene expression. If a mutation occurs within the *germ line*, it will occur in all the cells of an organism. Furthermore, a germ line mutation can be passed from parent to offspring. By comparison, *somatic* mutations only affect a patch of the organism. The size of the patch depends on how early during development the mutation has occurred. Mutations are random events, although *hot spots* for higher rates of mutation can occur at certain locations within genes. A variety of agents, known as *mutagens*, are known to induce mutations. These include many different types of *chemical* and *physical* agents. In addition, *spontaneous mutations* can arise during normal cellular conditions. Mutagens can alter DNA structure by modifying bases, removing bases, causing errors in DNA replication, and producing breaks in DNA strands. Testing methods such as the *Ames test* can determine whether an agent is mutagenic.

Due to the prevalence of mutagens in the environment, living cells have evolved DNA repair systems that can recognize and repair different types of DNA lesions. A few types of DNA repair systems can recognize altered bases and repair them directly. Alternatively, *nucleotide excision repair* systems recognize alterations in DNA structure and excise the abnormal region. Excision repair mechanisms are found in all organisms and represent a prominent, general system for DNA repair. Excision repair is particularly important in repairing actively transcribing genes. Certain proteins, such as the *transcription—repair coupling factor* of *E. coli*, can target the excision repair system to a damaged gene that is being actively transcribed. Finally, living organisms also possess *mismatch repair* and *recombinational repair* systems that operate after or during the DNA replication process.

Experimentally, there are many ways to study the occurrence, causes, and con sequences of mutation. Luria and Delbruck conducted a fluctuation test to show that mutations occur randomly in a population. Similarly, the Lederbergs' used replica plating to demonstrate that mutations occur randomly, and that a selective agent such as T1 phage simply selects for the growth of organisms that happen to have incurred a mutation providing T1 resistance. However, there are hot spots where mutations may occur more frequently. The work of Benzer showed that a few sites within two bacteriophage genes were more likely to mutate than were other sites.

Muller was the first scientist to establish that environment agents such as X-rays can cause mutation. His work showed that X-rays dramatically increased the likely hood of recessive, X-linked lethal alleles. Since these studies, many mutagens have been discovered by biochemists, microbiologists, and geneticists. Testing methods, such as the Ames test, can ascertain whether or not an agent is a mutagen.

9

TRANSPOSITION

Genetic recombination is the process in which chromosomes are broken and then rejoined to form a new genetic combination, different from the original. A major category of genetic recombination is homologous recombination, which is an essential feature of all organisms. As its name suggests, *homologous recombination* occurs between DNA segments that are homologous to each other. This process enhances genetic diversity, helps to maintain genome integrity (i.e., DNA repair), and ensures the proper segregation of chromosomes.

In this chapter, we will also examine ways that non homologous segments of DNA may recombine with each other. During *site-specific recombination*, non-homologous DNA segments are recombined at specific sites. This type of re combination occurs within genes that encode antibody polypeptides and also occurs when certain viruses integrate their genomes into host cell DNA. Finally, we will end this chapter with a description of an unusual form of genetic re combination known as *transposition*. As we will learn, small segments of DNA called *transposons* can move themselves to multiple locations within the host's chromosomal DNA.

From a molecular viewpoint, homologous recombination, site-specific re combination, and transposition all are important mechanisms for DNA re arrangement. These processes involve a series of steps that direct the breakage and rejoining of DNA fragments. Various cellular proteins are necessary for these steps to occur properly. The past few decades have seen many exciting advances in our understanding of genetic recombination at the molecular level. In the first part of this chapter, we will consider the general concepts of recombination and examine several models that explain how recombination occurs.

Sister Chromatid Exchange and Homologous Recombination

As discussed already that chromosomes have similar or identical sequences frequently participate in crossing over during meiosis I and occasionally during mitosis. As you may recall, crossing over involves the alignment of a pair of homologous chromosomes, followed by the breakage of two chromosomes at analogous locations, and the subsequent exchange of the corresponding segments. When crossing over occurs between sister chromatids, it is called *sister chromatid exchange* (SCE). Since sister chromatids are genetically identical to each other, SCE does not produce a new combination of alleles. There fore, it is not considered a form of recombination. By comparison, it is also common for homologous chromosomes to cross over. This is considered homologous recombination, because two similar (but not identical) homologues have exchanged genetic material. Homologous recombination may produce a new combination of alleles in the resulting chromosomes. *Recombinant chromosomes* contain a combination of alleles not found in the *parental chromosomes*.

In this section, we will begin with an experimental approach to detect sister chromatid exchange, overcoming the obstacle that the exchanged chromosomes are genetically identical to each other. We will then focus our attention on the molecular mechanisms that underlie homologous recombination.

Experiment

Perry and Wolff Produced Harlequin Chromosomes to Reveal Recombination between Sister Chromatids

Our understanding of crossing over and genetic recombination has come from a variety of experimental approaches including genetic, biochemical, and cytological analyses. Chromosomal staining methods have made it possible to visualize the genetic exchange between eukaryotic chromosomes. In the 1970s, the Russian cytogeneticist A. F. Zakharov and colleagues spent much effort developing methods that improved our ability to identify chromosomes. They made the interesting observation that chromosomes labeled with the nucleotide analogue 5-bromodeoxyuridine (BrdU) become more fluorescent when stained with Giemsa and then visualized microscopically. As we will see in the present experiment, Paul Perry at the University of California in San Francisco and Sheldon Wolff in Edinburgh, U.K., extended this approach to differentially stain sister chromatids and microscopically identify sister chromatid exchanges.

Before we consider the experiment of Perry and Wolff, let's examine how their staining procedure allowed them to accurately discern the two sister chromatids. In their approach, eukaryotic cells were grown in a laboratory and exposed to BrdU for two rounds of DNA replication. After the second round of DNA replication, one of the sister chromatids contained one normal strand and one BrdU-labeled strand. The other sister chromatid had two BrdU-labeled strands .When treated with two dyes, Hoechst 33258 and Giemsa, the sister chromatid containing two strands with BrdU stains very weakly and appears light, whereas the sister chromatid with only one strand containing BrdU stains much more strongly and appears very dark. In this way, the two sister chromatids can be distinguished microscopically. Chromosomes stained in this way have been referred to as ***harlequin chromosomes***, because they are reminiscent of a harlequin character's costume with its variegated pattern of light and dark patches.

Hypothesis

A chromosomal staining method will make it possible to identify recombination events where genetically identical pieces of sister chromatids are exchanged with each other.

Starting material: A laboratory cell line of Chinese hamster ovary (CHO) cells.

1. Expose CHO cells to BrdU for two cell generations (approximately 24 hours).
2. Near the end of the growth, expose the cells to colcemid. This prevents the cells from completing mitosis following the second round of DNA replication.
3. Add 0.075 M KCl to spread the chromosomes and then methanol/ acetic acid to fix the cells.
4. Stain with Hoechst 33258, rinse, and later stain with Giemsa. Note: This refinement in the staining procedure greatly improved the ability to discern the sister chromatids.
5. View under a microscope.

Interpreting the data

In this study, Perry and Wolff found that SCEs occurred at a frequency of approximately 0.67 per chromosome. This method has provided an accurate (and dramatic) way to visualize genetic exchange between eukaryotic chromosomes.

Many subsequent studies have utilized the harlequin staining method to study the effects of agents that may influence the frequency of genetic exchanges. Researchers have found that DNA damage caused by radiation and chemical mutagens tends to increase the level of genetic exchange. When cells are exposed to these types of mutagens, the technique of harlequin staining has revealed a substantial increase in the frequency of SCEs.

Holiday Model Describes a Molecular Mechanism for the Recombination Process

We will now turn our attention to genetic exchange that occurs between homologous chromosomes. Perhaps it is surprising that the first molecular model of homologous recombination did not come from a biochemical analysis of DNA or from electron microscopy studies. Instead, it was deduced from the outcome of genetic crosses in fungi.

As discussed already that geneticists have learned a great deal from the analysis of fungal asci, because an ascus contains the products of a single meiosis. When two haploid fungi that differ at a single gene are crossed to each other, it is expected that the ascus will contain an equal proportion of each genotype. For example, if a pigmented strain of *Neurospora* producing orange spores is crossed to an albino strain producing white spores, the octad should contain four orange spores and four white spores. As early as 1934, however, H. Zickler noticed that unequal proportions of the spores sometimes occurred within asci. He occasionally observed octads with six orange spores and two white spores, or six white spores and two orange spores.

Zickler used the term *gene conversion* to describe this phenomenon. It occurred at too high a rate to be explained by new mutations. Subsequent studies in the 1950s by several researchers confirmed this phenomenon in yeast and *Neurospora*. When gene conversion occurs, one allele is converted to the allele on the homologous chromosome.

Based on studies involving gene conversion, Robin Holliday at the National Institute for Medical Research in London proposed a model in 1964 to explain the molecular steps that occur during homologous recombination. In this chapter, we will first consider the steps in the Holiday model and then move on to more recent models. Later in this chapter, we will examine how the Holliday model can explain the phenomenon of gene conversion.

At the beginning of the process depicted in this figure, two homologous chromosomes are aligned with each other. According to the Holiday model, in step 1 a break occurs at identical sites in one strand of both parental chromosomes. During step 2, the strands then invade the opposite helices and base pair with the complementary strands. In step 3, this event is followed by a covalent linkage to create a *Holliday junction*.

In step 4, the cross in the Holiday junction can migrate in a lateral direction. As it does so, a DNA strand in one helix is being swapped for a DNA strand in the other helix. This process is called *branch migration*, because the branch connecting the two double helices migrates laterally. Since the DNA sequences in the homologous chromosomes are similar but not identical, the swap ping of the DNA strands during branch migration may produce regions in the double-stranded DNA that are called *heteroduplexes*. A heteroduplex is a DNA double helix that contains mismatches. In other words, since the DNA strands in this region are from homologous chromosomes, their sequences are not perfectly complementary, yielding mismatches.

During step 5, the Holiday structure may or may not make a 180° turn. This is called *isomerization*, because the two structures shown at step 5 are structural *isomers* of each other. This means that they are chemically identical except for the relative locations of certain segments.

The final steps in the recombination process are called *resolution*, since they involve the breakage and rejoining of two DNA strands to create two separate chromosomes. In other words, the entangled DNA strands become resolved into two separate structures. Resolution can occur in two ways. Steps 6A—8A are the result without isomerization, whereas steps 6B—8B will occur if isomerization has taken place at step 5. As shown in steps 6A—8A, breakage can occur in the same two strands that were also broken in step 1. If this occurs, the strands are rejoined to produce a non-recombinant pair of chromosomes. The only difference between this non-recombinant pair and the original parental pair is the short heteroduplex region. Alternatively, as shown in steps 6B—8B, the resolution phase can involve breakage of the two DNA strands that were not broken during step 1. In this case, the rejoining of the corresponding strands produces two recombinant chromosomes.

The Holiday model can account for the general properties of recombinant chromosomes that are formed during eukaryotic meiosis. As was mentioned, the original model was based on the results of crosses in fungi where the products of meiosis are contained within a single ascus. Nevertheless, molecular research in many other organisms have supported the central tenets of the Holliday model. A particularly convincing piece of evidence came from electron microscopy studies in which recombination structures could be visualized an electron micrograph of two DNA fragments that are in the process of recombination. This structure can be equated to those found at step 5. They have been referred to as chi (χ) forms because their shape is similar to the Greek letter χ.

More Recent Models have Refined the Molecular Steps of Homologous Recombination

As more detailed studies of genetic recombination have become available, certain steps in the Holiday model have been reconsidered. In particular, more recent models have modified the initiation phase of recombination. It is now known that the first step need not involve nicks at identical sites in two corresponding strands. A DNA helix with a break in both strands or DNA molecules with a single nick have been shown to participate in genetic recombination. Therefore, newer models have tried to incorporate these experimental observations. A model proposed by Matthew Meselson (Harvard University) and Charles Radding (Yale University) hypothesizes that a single nick in one DNA strand initiates recombination. A second model, pro posed by Jack Szostak, Terry Orr-Weaver, Rodney Rothstein, and Franldin Stahl, suggests that a double-stranded break initiates the recombination process. This is called the double-stranded break model.

A few important differences from the Holiday model are worth noting in these newer models. As mentioned, the pattern of breakage of one or more DNA strands is different among the three models. Unlike the Holiday model, the newer models propose that a short region of strand degradation occurs. This occurs via the action of nucleases that can degrade a DNA strand over a short distance. (This occurs in step 3 in the Meselson—Radding model and step 1 in the double-stranded break model.) Since DNA strand degradation takes place, these two models also re quire the synthesis of new DNA. This DNA synthesis occurs in the relatively short gaps where a DNA strand is missing. For this reason, the DNA synthesis is called *DNA gap repair synthesis*. The arrowheads and c indicate the areas where DNA gap repair synthesis occurs. In the Meselson—Radding model, gap repair synthesis occurs in one strand; in the double-stranded break model, it occurs in two strands.

Various Proteins are Necessary to Facilitate Homologous Recombination

The homologous recombination process requires the participation of many proteins that catalyze different steps in the recombination pathway. Though homologous recombination takes places in all organisms, the enzymology of this process is best understood in *Escherichia coli*. Figure presents a more complete version of the double-stranded break model that includes some of the *E. coli* proteins that play critical roles in this process. RecBCD is a multi-component protein composed of the RecB, RecC, and RecD proteins. (The term Rec indicates that these proteins are involved with recombination.) The RecBCD complex plays an important role in the initiation of recombination involving double-stranded breaks. Rec BCD recognizes a double-stranded break within DNA and catalyzes DNA unwinding and strand degradation. The action of RecBCD produces single-stranded DNA ends that can participate in strand invasion and exchange.

The RecA protein can bind to the single-stranded ends of DNA molecules that are generated from the activity of RecBCD. A large number of RecA proteins bind to single-stranded DNA, forming a structure called a filament. During *synapsis*, this filament makes contact with the unbroken chromosome. Initially, this contact is most likely to occur at non homologous regions. The contact point slides along the DNA until it reaches a homologous region. Some current models suggest that a homologous region is recognized by formation of triplex DNA, although this is not firmly established. One of the original strands is quickly displaced, and the invading single-stranded DNA forms a normal double helix with the other strand. RecA proteins mediate the movement of the invading strand and the displacement of the complementary strand. In step 3, this displaced strand invades the vacant region of the broken chromosome.

Proteins that bind specifically to Holliday junctions have also been identified. In the double-stranded break model, DNA gap repair synthesis creates a double Holliday junction. RecG and RuvAB proteins specifically bind to Holiday junctions. Either of these proteins can catalyze the branch migration of Holliday junctions. RuvC protein also recognizes Holiday junctions. It is called a *resolvase*, because it is an endonuclease that makes the final cuts in the DNA during the resolution phase of recombination.

Gene Conversion may Result from DNA Gap Repair Synthesis or DNA Mismatch Repair

As was mentioned earlier in the chapter, genetic recombination can lead to an event where two different alleles become two identical alleles. Since one of the alleles has been converted to the other, this process is known as gene conversion. The Holliday model, as well as newer models, can account for the phenomenon of gene conversion. There are two possible ways that gene conversion can occur. One way is via DNA gap repair synthesis. Figure illustrates how gap repair synthesis can lead to gene conversion according to the double-stranded break model. The top chromosome, which carries the *a* allele, has suffered a double-stranded break. A gap is created by the digestion of the DNA in the double helix. This digestion eliminates the a allele. The two template strands used in gap repair synthesis are from the other double helix. This helix carries the A allele. After gap repair synthesis takes place, the top chromosome will contain the *A* allele.

A second mechanism to account for gene conversion is DNA mismatch repair. To understand how this works, let's take a closer look at the heteroduplex structure that is formed during homologous recombination. A heteroduplex contains a DNA strand from each of the two original parental chromosomes. It is possible that the two parental chromosomes may contain an allelic difference within this region. In other words, this short region may contain DNA sequence differences. If this is the case, the heteroduplex region that is formed after branch migration will contain an area of base mismatch. Gene conversion occurs when recombinant chromosomes are repaired to produce the same allele.

The two parental chromosomes contained an allelic difference in their DNA sequences as shown at the top of the figure. During recombination, branch migration has occurred across this region, thereby creating two heteroduplexes with base mismatches. DNA mismatches will be recognized by DNA repair systems and repaired to a DNA double helix that obeys the A—T/G—C rule. These two mismatches can be repaired in four possible ways. As shown here, two possibilities will produce no gene conversion, whereas the other two will lead to gene conversion.

Site-specific Recombination

Thus far in this chapter, we have examined recombination between segments of DNA that are homologous or identical to each other. Site-specific recombination is another mechanism where DNA fragments can recombine to make new genetic combinations. During this type of recombination, two DNA segments with little or no homology align themselves at specific sites. The sites are relatively short DNA sequences (a dozen or so nucleotides in length) that provide a specific location where recombination will occur. Chromosome breakage and reunion occurs at these defined sites to create a recombinant chromosome. These sites are recognized by specialized enzymes that catalyze the breakage and rejoining of DNA fragments within the sites.

Certain viruses use site-specific recombination to insert their viral chromosome into their host cell's chromosome. This process has been examined carefully in bacteriophage λ. In addition, mammalian genes that encode antibody polypeptides are rearranged by site-specific recombination, which enables the generation of a diverse array of antibodies. In this section, we will consider both types of mechanisms.

Integration of Viral Genomes can Occur by Site-specific Recombination

The life cycle of some viruses involves the integration of viral DNA into host cell DNA. Certain bacteriophages, for example, can integrate their viral DNA into the bacterial chromosome, creating a *prophage*. This prophage can exist in a latent, or *lysogenic*, state for many generations. The integration of phage DNA is well understood for bacteriophage λ, which infects *E. coli*. The integration occurs by a mechanism involving site-specific recombination.

Integration of the λ DNA into the *E. coli* chromosome requires sequences known as *attachment sites*. As shown at the top, the attachment site sequences are identical in the λ DNA and the *E. coli* chromosome. An enzyme known as *integrase* is encoded by a gene in the λ DNA. Several molecules of integrase recognize the attachment site sequences and bring them close together. Integrase then makes staggered cuts in both the λ and *E. coli* attachment sites. The strands are then exchanged, and the ends are ligated together. In this way, the phage DNA is integrated into the host cell chromosome. As a prophage, the λ DNA may remain latent for many generations. Certain conditions (e.g., when the bacterium is exposed to UV light) may act to stimulate the excision of the prophage from the host DNA, thereby reactivating the virus. Excision also requires integrase, which catalyzes the reverse reaction, as well as a second protein known as *excisionase*.

Antibody Diversity in the Immune System is Produced by Site-specific Recombination

As we have just seen, viruses can integrate their DNA into the host cell chromosome by a site-specific recombinational event. This process requires a viral enzyme, integrase, that recognizes the sites and catalyzes the recombination reaction. A similar process occurs in certain cells of the immune system. The DNA sequences within antibody genes are rearranged by enzymes that recognize specific sites within those genes and catalyze the breakage and reunion of DNA segments. Before we discuss the details of this mechanism, it is interesting to consider the biology of antibodies.

Antibodies or immunoglobulins (Igs) are proteins produced by the B-cells of the immune system. Their function is to recognize foreign substances (namely, viruses, bacteria, etc.) and target them for destruction. Antibodies recognize sites known as epitopes within the structures of foreign substances (also known as antigens). The recognition between an antibody and antigen is very specific, with each type of antibody recognizing a single epitope. Within the immune system, each B-cell produces a single type of antibody. However, our bodies have millions of B-cells. Site-specific recombination allows each B-cell to produce an antibody with a different amino acid sequence. These differences in the amino acid sequences of antibody proteins enable them to recognize different epitopes. In this way, the immune system can identify an impressive variety of substances as being foreign antigens and thereby target them for destruction.

From a genetic viewpoint, the production of millions of different antibodies poses an interesting problem. If a distinct gene were needed to produce each different antibody polypeptide, the DNA would need to contain millions of different antibody genes. By comparison, consider that the entire human genome contains only about 100,000 different genes. To generate millions of antibody molecules with different polypeptide sequences, an unusual mechanism has evolved in which DNA is cut and reconnected by site-specific recombination. We might call this "DNA splicing," although the term splicing is usually reserved for cutting and re joining of RNA molecules. With this mechanism, only a few large antibody pre cursor genes are needed to produce millions of different antibodies. These precursor genes are spliced in many different ways to produce a vast array of polypeptides with differing amino acid sequences.

Antibodies are tetrameric proteins composed of two heavy polypeptide chains and two light chains. One type of light chain is the κ (kappa) light chain, which is a component of a mass of antibodies known as immunoglobulin G (IgG). At the left side, there are 300 regions known as variable (V) sequences or domains. In addition, there are four different joining (J) sequences and a single constant (C) sequence. Each variable domain or joining domain encodes a different amino acid sequence.

During the differentiation of B-cells, the κ light precursor gene is cut and re joined so that one variable sequence becomes adjacent to a joining sequence. This recombination event is catalyzed by an enzyme known as *V(D)J recombinase*. At the 3' end of every V domain and the 5' end of the J domains is located a *recombination signal sequence* that functions as a location for site-specific recombination between the V and J regions. These signal sequences are recognized by a V(D)J recombinase that carries out the recombination reaction. Following transcription, the fused V region is contained within a pre-mRNA transcript that is also spliced to connect the J and C regions. After this has occurred, a B-cell will only produce the particular κ light chain encoded by the specific VJ fusion domain and the constant domain.

The recombination process within immunoglobulin genes produces an enormous diversity in polypeptides. Even though it occurs at specific junctions within the antibody gene, the recombination is random with regard to the particular V and I segments that can be joined. Any of the 300 different variable sequences can be spliced next to any of the four joining sequences. This amounts to 1200 possible combinations.

The heavy-chain polypeptides are produced by a similar recombination mechanism. In this case, there are about 500 variable segments and four joining segments. In addition, there are also 12 diversity (D) segments, which are found between the variable and joining segments. The recombination first involves the connection of a D and J segment, followed by the connection of a V and DJ segment. The number of heavy-chain possibilities is 500 × 12 × 4 = 24,000. Since any light chain—heavy chain combination is possible, this yields 1200 × 24,000 = 28,800,000 possible anti body molecules from the splicing of only two precursor genes!

Transposition

The last form of recombination that we will discuss is transposition. In some ways, transposition resembles the site-specific recombination that we examined for phage λ. In that case, a segment of λ DNA was able to integrate itself into the *E. coli* chromosome. Transposition also involves the integration of small segments of DNA into the chromosome. Transposition, though, can occur at many different locations within the genome. The DNA segments that transpose themselves are known as *transposable elements* (TEs). TEs have sometimes been referred to as "jumping genes," because they are inherently mobile.

Transposable elements were first identified by Barbara McClintock in the early 1950s from her classical studies with corn plants. Since that time, geneticists have discovered many different types of TEs in species as diverse as bacteria, fungi, plants, and animals. The advent of molecular technology has allowed scientists to understand more about the characteristics of TEs that enable them to be mobile. In this section, we will examine the characteristics of transposable elements and explore mechanisms that explain how TEs move. We will also discuss the biological significance of TEs and their uses as experimental tools.

Mc Ctintock found that chromosomes of corn plants contain loci that can move

Barbara McClintock began her scientific career as a student at Cornell University. Her interests quickly became focused on the structure and function of the chromosomes of corn plants, an interest that continued for the rest of her life. She spent countless hours examining corn chromosomes under the microscope. She was technically gifted and, in addition, had a theoretical mind that could propose ideas that conflicted with conventional wisdom.

During her long career as a scientist, McClintock identified many unusual features of corn chromosomes. She noticed that one strain of corn had the strange characteristic that a particular chromosome, number 9, tended to break at a fairly high rate at the same site. McClintock termed this a *mutable site* or *locus*. This observation initiated a six-year study concerned with highly unstable chromosomal locations. In 1951, at the end of her study, McClintock proposed that these mutable sites are actually locations where transposable elements have been inserted into the chromosomes. At the time of McClintock's studies, such an idea was entirely unorthodox.

McClintock focused her efforts on the relationship between a mutable locus and its phenotypic effects on corn kernels. There were several genes on a single chromosome that could be used as chromosomal markers in her crosses. All of these genes affected the phenotype of corn kernels. Each gene had (at least) one dominant and one recessive allele that could be detected easily by examining the kernels. A few of these alleles are described here. The chromosome also contained a mutable locus that McClintock termed *Ds* (for dissociation) since the locus was known to frequently cause chromosomal breaks. In the chromosome shown here, the Ds locus is located next to several genes affecting kernel traits.

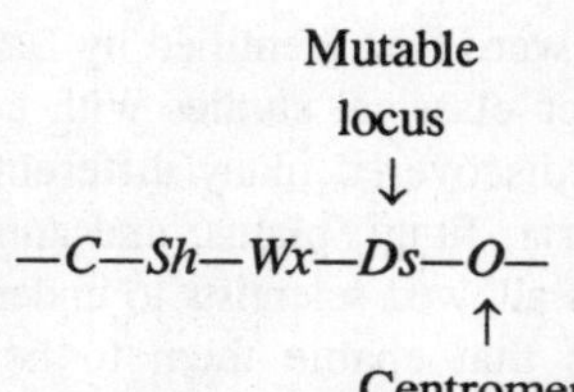

C = normal kernel color (dark red); c = colorless kernel; C^I = a dominant allele that also causes colorless kernel

Sh = normal endosperm; sh = shrunken endosperm

Note: The endosperm is the storage material in the kernel that is used by the plant embryo to provide energy for growth.

Wx = normal starch in the endosperm; wx = (waxy), no starch in the endosperm

McClintock conducted crosses demonstrating that the *Ds* locus could be transposed to different locations within the corn genome. Using the chromosome shown here, it was straightforward to determine if *Ds* had moved, because the movement of *Ds* occasionally causes chromosome breakage. Before discussing her crosses, it important to mention that the endosperm of a kernel is triploid: it is derived from two maternal haploid nuclei and one paternal (i.e., pollen) haploid nucleus. McClintock produced a cross with the following genotype:

_____ *C^I* __ *Sh* ____ *Wx* __ *Ds* ________ O _______
_____ *C* __ sh ____ *wx* ______________ O _______
_____ *C* __ *sh* ____ *wx* ______________ O _______

This kernel is expected to be colorless, because the *C^I* allele is dominant and causes a colorless phenotype. However, as the kernel develops, some chromosomes may break at the *Ds* locus and lose the distal part of this chromosome:

________ O _______
_____ *C* __ *sh* ____ *wx* ______________ O _______
_____ *C* __ *sh* ____ *wx* ______________ O _______

Within a single kernel, this will produce a patch of cells having a different phenotype. These patches will be red, waxy, and shrunken.

By analyzing many kernels, McClintock was able to identify cases in which *Ds* had moved to a new location. For example, if *Ds* had inserted between *Sh* and *Wx*, a break at *Ds* would produce the following combination:

_____ *Wx* _______________ O ________

______ *C* __ *sh* _____ *wx* _______________ O ________

______ *C* __ *sh* _____ *wx* _______________ O ________

This genotype would produce patches on the kernel that are red and shrunken but not waxy. In this way, McClintock identified 20 independent cases in which the *Ds* element had moved to a new location within this chromosome. Overall, the results from many crosses were consistent with the idea that *Ds* can transpose itself throughout the corn genome.

McClintock also found that a second locus, termed *Ac* (for activator), was necessary for *Ds* transposition to occur. It is now known that the *Ac* locus contains a gene that encodes the enzyme transposase, which is necessary for the *Ds* region to move. We will discuss the function of transposase later in this chapter. Some strains of McClintock's corn contained this activator locus, whereas others did not.

In one cross, McClintock noticed a particularly exciting and unusual event. She started with a plant that had the following genotype:

______ *C* __ *Sh* ____ *wx* ___ *Ds* _________ O ________

______ *c* ___ *sh* _____ *wx* _______________ O ________

______ *c* ___ *sh* _____ *wx* _______________ O ________

In this cross, the kernels are expected to be red. If the strain contains the *Ac* locus, breakage will occur occasionally at the *Ds* locus to produce colorless patches. Among 4000 kernels, she noticed one kernel with the opposite phenotype. It had a colorless background with red patches. This suggested that the background genotype was recessive, *c*, and it was mutable to become *C*! In this kernel, the *Ds* element had moved into the *C* gene:

____ *CDsC* _ *Sh* ____ *wx* _______________ O ________

______ *c* ___ *sh* _____ *wx* _______________ O ________

______ *c* ___ *sh* _____ *wx* _______________ O ________

Furthermore, the red patch phenotype required the *Ac* locus.

McCkntock postulated that the colorless phenotype was due to a transposition of *Ds* into the *C* gene. When *Ds* is located within the *C* gene, she proposed, it inactivated the *C* gene, producing the recessive colorless phenotype. However, when *Ds* occasionally transposed out of the gene, the *C* allele would be restored and a red patch would result. (Note: In this case, the formation of patches, or sectoring, is due to the movement of *Ds* out of its original location but not due to chromosome breakage.) According to this hypothesis, the red phenotype should be associated with two observations. First, *Ds* should have moved to a new location. Second, the restored *C* allele should no longer be mutable.

Hypothesis

The transposition of the *Ds* element into the normal C gene prevents kernel pigmentation. When the *Ds* element transposes back out of the *C* gene, the normal *C* allele is restored and the kernel becomes red.

Testing the hypothesis

Starting material: The male pollen of the corn plant had a chromosome in which the *Ds* element had moved into the *C* gene. The male plant also contains the *Ac* locus.

1. Cross the male with a female that is *c*, *sh*, *Wx*, (no *Ds*) homozygous.
2. In this cross, most kernels will have a white background with red sectoring. On rare occasions, *Ds* may have transposed during male gamete formation, producing a completely red kernel.
3. Identify those occasional kernels that are completely red.
4. Germinate the solid red kernels.
5. Conduct crosses to determine the location of *Ds* in the plants derived from the solid red kernels. Note: This can be done using the chromosomal markers and the strategy that was described at the beginning of this experiment. When observing the results of these crosses, determine whether the *C* gene is still mutable. (Is red sectoring occurring?)

Interrupting the data

In the parent strain, previous results showed that the *Ds* locus had transposed into the *C* gene. This parent strain had red sectoring on a colorless background. This result is consistent with the idea that the red sectoring is due to the removal of the *Ds* locus from the *C* gene, thereby restoring a normal *C* gene. Solid red kernels occasionally arose from this parental strain. By conducting the appropriate crosses, McClintock found that in the progeny of the red kernels, the *Ds* locus had moved out of the *C* gene to another location. In addition, the "restored" *C* gene behaved normally. In other words, it was no longer a highly mutable gene exhibiting a sectoring phenotype. Taken together, the results are consistent with the idea that the *Ds* locus can move around the corn genome by transposition.

When McClintock published these results in 1951, they were met with great skepticism. Most geneticists of this time were unable to accept the idea that the genetic material was susceptible to frequent rearrangement. Instead, they wanted to believe that the genetic material was always very stable and permanent in its structure. It was not until several decades later that the scientific community accepted the existence of transposable elements. Much like Gregor Mendel and Charles Darwin, Barbara McClintock was clearly ahead of her time. She was awarded the Nobel prize in 1983, more than 30 years after her original discovery.

Transposable Elements and Retroelements Move via One of Three Transposition Pathways

Since the pioneering studies of Barbara McClintock, many transposable elements have been found in bacteria, fungi, plant, and animal cells. Three general types of transposition pathways have been identified. In *simple* or *conservative transposition*, the TE is removed from its original site and transferred to a new target site. By Comparison, *replicative transposition* involves the replication of the TE and insertion of the newly made copy into a second site. In this case, one of the TEs remains in its original location and the other is inserted at another location. A third category of elements move via an RNA intermediate. These are known as *retroelements*, *retrotransposons*, or *retroposons*. Like replicative transposition, retroelements increase in number during retrotransposition events.

Each Type of Transposable Element has a Characteristic Pattern of DNA Sequences

As researchers have studied TEs from many species, they have found that DNA sequences within transposable elements are organized in several different ways. A few ways that TEs are organized, although many variations are possible. The simplest TEs, which are commonly found in bacteria, are known as *insertion sequences*. Insertion sequences have two important characteristics. First, both ends of the insertion sequence contain *inverted repeats* (IRs). Inverted repeats are DNA sequences that are identical (or very similar) but run in opposite directions, such as the following:

5'-CTGACTCTT-3' and 5'-AAGAGTCAG-3'
3 '-GACTGAGAA-5' 3'-TTCTCAGTC-5'

In addition, insertion sequences may contain a central region that encodes the enzyme *transposase*, which catalyzes the transposition event.

In the case of replicative transposons, the enzyme resolvase is also encoded.

Composite transposons contain additional genes that are not necessary for transposition *per se*. Composite transposons commonly contain genes that confer a selective advantage to the organism. Composite transposons are prevalent in bacteria, where they often contain genes that provide resistance to antibiotics or toxic heavy metals. Composite transposons contain such a resistance gene that is sandwiched between two TEs. For example, the composite transposon contains two insertion sequences flanking a gene that confers antibiotic resistance. During transposition of a composite transposon, only the two inverted repeats at the ends of the transposon are involved in the transpositional event. Any time insertion sequences are found at both ends of a gene, they create a composite transposon.

The organization of retroelements can be quite variable, and they are categorized based on their evolutionary relationship to retroviral sequences. Retroviruses are RNA viruses that make a DNA copy that integrates into the host's genome. The viral-like retroelements contain *long terminal repeat sequences* (LTRs) at both ends of the element. These elements tend to be fairly large and show striking similarities to known retroviruses. In particular, viral-like retroelements encode virally related proteins such as *reverse transcriptase* and integrase.

By comparison, non-viral-like retroelements appear less like retroviruses in their sequence, although some similarity, such as the occurrence of a reverse transcriptase gene, may be present. Many non-viral-like retroelements, however, do not share any sequence similarity with known viruses. Instead, some non-viral-like retroelements are evolutionarily derived from normal eukaryotic genes. For example, the Alu family of repetitive sequences found in humans is derived from a single ancestral gene known as the 7SL RNA gene. This gene sequence has been copied by retroposition to achieve the current number of greater than 500,000 copies. Incredibly, it constitutes approximately 5 to 6% of the human genome!

Transposable elements are considered to be *complete* (or *autonomous*) *elements* when they contain all the information necessary for transposition or retroposition to take place. However, TEs are often *incomplete* (or *nonautonomous*). An incomplete element typically lacks a gene such as transposase or reverse transcriptase that is necessary for transposition. The *Ds* element is an incomplete element that lacks a transposase gene. The *Ac* locus provides a transposase gene that enables *Ds* to transpose. Therefore, incomplete TEs such as *Ds* can trans pose when the transposase gene is present at another region in the genome.

Transposase Catalyzes the Excision and Insertion of Transposable Elements

Now that we have an understanding of the sequence organization of transposable elements, we can examine the steps of the transposition process. The enzyme transposase catalyzes the removal of a TE from its original site in the chromosome and its subsequent insertion at another location. Transposase proteins recognize the inverted repeat sequences at the ends of the TE and bring them close together. The DNA is cleaved at the ends of the TE, excising it from its original site within the chromosome. This excision event is followed by an insertion process in which transposase cleaves the target DNA sequence at staggered recognition sites. The TE is then ligated to the target DNA.

The ligation of the transposable element into its new site initially leaves short gaps in the target DNA. Notice that the DNA sequences in these gaps are complementary to each other (in this case, ATGCT and TACGA). Therefore, when they are filled in by DNA gap repair synthesis, the DNA base pair sequences at both ends of the TE are identical. These two sequences are called *direct repeats* (DRs), because they are in the same *direction* and they are *repeated* at both ends of the element. The occurrence of direct repeats is a common feature of all TEs.

Replicative Transposition Requires both Transposase and Resolvase

Replicative transposition has been studied in several bacterial transposons and in bacteriophage (mu), which behaves like a transposon. The net result of replicative transposition is that a transposable element occurs at a new site and the original TE remains in its original location. One DNA molecule already has a TE, whereas the other does not. In this mechanism, transposase initially makes two cuts in the TE and in the target DNA. Note that this differs from conservative transposition, in which the transposase makes four cuts in the DNA and thereby removes the TE from its original site. In replicative transposition, one strand of the TE is left at its original location.

Following ligation, both the host DNA and the transposable element have a long gap. Gap repair DNA synthesis copies the host DNA gap as well as the TE. This creates two copies of the TE within a large circular molecule known as a *cointegrant*. The enzyme resolvase catalyzes homologous recombination within the TEs so that the cointegrant can be resolved into two separate DNA molecules. One of these molecules contains the TE in its original location, and the other has a TE at a new location.

In the preceding discussion, we have seen that replicative transposons are duplicated when they transpose, whereas conservative TEs are simply removed and placed in a new location. Thus, you might think that conservative TEs would have difficulty increasing in number. Interestingly, this is not the case. Conservative TEs can easily increase in number, because transposition often occurs during DNA replication. For conservative TEs, an increase in number can happen in the following way. After a replication fork has passed a region containing a TE, there will be two TEs behind the fork (i.e., one in each of the replicated regions). One of these TEs could then trans pose from its original location into a region ahead of the replication fork. After the replication fork has passed this second region and DNA replication is completed, there will be two TEs in one of the chromosomes, and one TE in the other chromo some. In this way, conservative TEs can increase in number. We will discuss the biological significance of transposon proliferation later in this chapter.

Retroelements Utilize Reverse Transcriptase and Integrase for Retrotransposition

Thus far, we have considered how DNA elements (i.e., transposons) can move throughout the genome. By comparison, retroelements utilize an RNA intermediate in their transposition mechanism. The movement: of retroelements also requires two key enzymes, reverse transcriptase and integrase. In this example, the host cell already contains a retroelement known as the Alu sequence within its genome. This retroelement, which behaves like a gene, is transcribed into RNA. Reverse transcriptase uses this RNA as a template to synthesize a complementary strand of DNA. The single-stranded DNA is then used as template by DNA polymerase to make double-stranded DNA. The ends of the double-stranded DNA are then recognized by integrase, which catalyzes the insertion of the DNA into the host chromosomal DNA. The integration of retroelements can occur at many locations within the genome. Furthermore, since a single retroelement can be copied into many RNA transcripts, retroelements may accumulate rapidly within a genome.

Transposable Elements may have Important Influences on Mutation and Evolution

Over the past few decades, researchers have found that transposable elements probably occur in the genomes of all species. The genomes of eukaryotic species typically contain moderately and highly repetitive sequences. In some cases, these repetitive sequences are due to the proliferation of TEs. In mammals, for example, *LINEs* are long interspersed elements that are usually I to 5 kbp in length and found in 20,000 to 100,000 copies per genome. *SINEs* are short interspersed elements that are less than 500 bp in length. A specific example of a SINE is the Alu sequence, present in 500,000 to 1,000,000 copies in the human genome. The biological significance of transposons in the evolution of prokaryotic and eukaryotic species remains a matter of intense debate. According to the *selfish DNA theory*, TEs exist because they contain the characteristics that allow them to multiply within the host cell DNA. In other words, they resemble parasites in the sense that they inhabit the host without offering any selective advantage. They can proliferate within the host as long as they do not harm the host to the extent that they significantly disrupt survival.

Alternatively, other geneticists have argued that most transpositional events are deleterious and, therefore, TEs would be eliminated from the genome if they did not also offer a compensating advantage. Several potential advantages have been suggested. For example, TEs may cause greater genetic variability by promoting recombination. In addition, bacterial TEs often carry an antibiotic resistance gene that provides the organism with a survival advantage. Researchers have also suggested that transposition may cause the insertion of exons into the coding sequences of structural genes. This phenomenon, called *exon shuffling*, may lead to the evolution of genes with more diverse functions.

While this controversy remains unresolved, it is clear that transposable elements can rapidly enter the genome of an organism and proliferate quickly. In *Drosophila melanogaster*, for example, a TE known as the P element likely was introduced into this species in the 1950s. Laboratory stocks of *D. melanogaster* collected prior to this time do not contain P elements. Remarkably, in the last 50 years, the P element has expanded throughout *D. melanogaster* populations world wide. The only strains without the P element are laboratory strains collected prior to the 1950s. This observation underscores the surprising ability of TEs to infiltrate a population of organisms.

Transposable elements have a variety of effects on chromosome structure and gene expression. Since many of these outcomes are likely to be harmful, transposition is usually a highly regulated phenomenon that only occurs in a few individuals under certain conditions. Agents such as radiation, chemical mutagens, and hormones stimulate the movement of TEs. When it is not carefully regulated, transposition is likely to be potently detrimental. For example, in *D. melanogaster*, if M strain females (which lack P elements) are crossed with P strain males (which contain numerous P elements), the hybrid offspring contain a variety of abnormalities, which include a high rate of mutation and chromosome breakage. This deleterious outcome, which is called *hybrid dysgenesis*, occurs because the P elements can transpose freely.

Transposons have become Important Tools in Molecular Biology

The unique and unusual features of transposons have made them an important experimental tool in molecular biology. For example, the introduction of a transposon into a cell is a convenient way to abolish the expression of particular genes. If a transposon "hops" into a gene, it is likely to inactivate the gene's function. This phenomenon can be utilized to clone a particular gene in an approach known as transposon tagging. In this strategy, researchers use transposons in an attempt to clone novel genes.

The first example of transposon tagging involved an X-linked gene in *Drosophila* that affects eye color. *Drosophila* X-linked gene can exist in the wild-type (red) allele and a loss of function allele that causes a white-eyed phenotype. In 1981, Paul Bingham at Research Triangle Park in North Carolina, in collaboration with Robert Levis and Gerald Rubin at the Carnegie Institution of Washington, used transposon tagging to clone this gene.

Prior to their cloning work, a wild-type strain of *Drosophila* had been characterized that carried a transposable element called *copia*. From this red-eyed strain, a white-eyed strain was obtained in which the *copia* element had transposed to a region on the X-chromosome that corresponded to where the eye-color gene mapped. The researchers reasoned that the white-eyed phenotype could be due to the insertion of the *copia* element into the wild-type gene, thereby inactivating it.

To clone the eye-color gene, chromosomal DNA from this white-eyed strain was isolated, digested with restriction enzymes, and cloned into viral vectors. A DNA library is a collection of vectors that contain different pieces of chromosomal DNA. If a transposon has "jumped" into the eye-color gene, vectors that contain this gene will also contain the transposon sequence. In other words, the presence of the transposon tags the eye-color gene. Therefore, a radiolabeled fragment of DNA that is complementary to the transposon sequence can be used as a probe to identify plaques that also contain the eye-color gene. In the example of the method of transposon tagging was successful at cloning an eye-color gene in *Drosophila*.

Concluding Remark

Genetic recombination occurs when segments of DNA are broken and reconnected to form new combinations. During *homologous recombination*, homologous DNA regions become aligned and exchange DNA segments. This enhances genetic variability by producing *recombinant chromosomes* with new combinations of alleles.

At the molecular level, several models have attempted to explain the steps in homologous recombination. The Holliday model was the first example of a molecular explanation for the recombination process. This model was able to account for the phenomenon of *gene conversion*, in which one allele of a gene is converted to another allele. More recently, other models have more accurately described certain steps in the recombination pathway. In addition, much progress has been made toward identifying the many proteins that play important roles in recombination.

A second mechanism for recombination is known as *site-specific recombination*, because the breakage and rejoining of the DNA segments occurs at particular DNA sequences. Site-specific recombination is responsible for the integration of certain bacteriophages such as λ into the host genome. In addition, site-specific recombination within *immunoglobulin* genes is important in generating an astounding diversity in *antibody* polypeptides. During this process, regions in precursor antibody genes known as V, D, and J are connected to each other via recombination. This mechanism enables the production of a very diverse array of antibody proteins.

Transposition is a third way that DNA segments can rearrange. Short segments of DNA known as *transposable elements* (TEs) possess characteristics that enable them to be mobile. During *conservative* transposition, a "cut and paste" mechanism occurs where the TE is cut out of its original site and ligated to a new site. *Transposase* catalyzes this reaction. By comparison, in *replicative* transposition a TE is duplicated, with one TE remaining in its original location and a new TE located at a new site. Finally, retroposition occurs via an RNA intermediate. In this case, the retroelement is transcribed to

RNA and then copied to DNA by *reverse transcriptase*. This DNA is then integrated into the chromosome by integrase. From an evolutionary view point, all mechanisms of transposition can cause many different types of mutations.

Crossing over, which may occur be sister chromatids or homologous chromosomes, can be detected by several techniques. As described already, sister chromatid exchange can be observed using staining methods that distinguish the sister chromatids within a pair. Homologous recombination can also be identified from the outcome of dihybrid crosses involving linked genes, where the frequency of recombinant offspring is a measure of the frequency of crossing over. At the molecular level, researchers have studied homologous recombination by identifying and characterizing the proteins and intermediates that facilitate the process. Holliday junctions are a type of intermediate involved in homologous recombination.

Likewise, researchers have elucidated the mechanisms of site-specific recombination and transposition using genetic and molecular techniques. In experiment, we learned that McClintock followed genetic crosses in corn to provide compelling evidence for the existence of transposable elements. Since that time, researchers have studied many different TEs. The sequencing of TEs has revealed certain patterns of DNA sequences that are required for transposition and retroposition. Researchers have also identified proteins that are necessary to pro mote the movement of TEs. In addition, molecular biologists have taken advantage of our knowledge of transposition and now routinely use transposons to inactivate and clone genes via transposon tagging.

10

Bacterial Mortality

Bacteriophages, or phages, have played an important role in the development of molecular biology. At the present time a few phages are the most completely understood of any organisms. Because of their lesser complexity than bacteria and higher cells and the availability of an enormous number of mutants, phages have been extraordinarily useful in the study of basic processes, such as replication, transcription, translation, and regulation. In a short book it is not possible to look into the molecular biology of phages in any depth. Instead, we shall examine some basic features of phage biology and provide a few examples of reproductive strategies by looking at particular stages of the life cycles of certain phages. For more information, consult MB.

A bacteriophage is, first of all, a bacterial parasite. By itself, it can persist, but a phage can neither grow nor replicate except within a bacterial cell. Most phages possess genes encoding a variety of proteins. However, all known phages use the ribosomes, protein-synthesizing factors, amino acids, and energy-generating systems of the host cell.

Each phage must perform some minimal functions for continued survival. These are the following:

1. To protect its nucleic acid from environmental chemicals that could alter the molecule.
2. To deliver its nucleic acid to the inside of bacterium.
3. To convert an infected bacterium to a phage-producing system which yields a large number of progeny phage.
4. To release phage progeny from an infected bacterium.

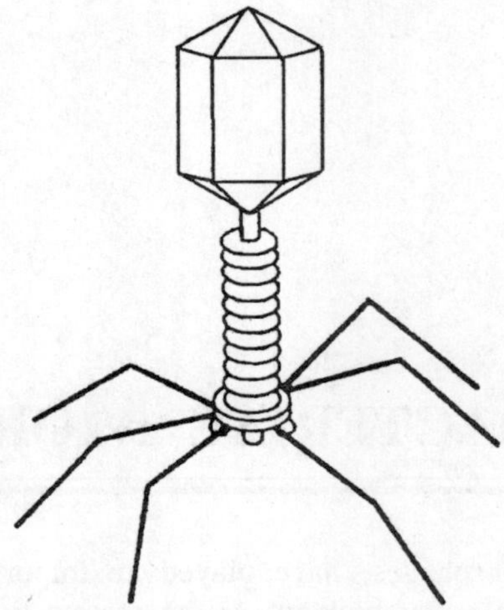

Fig. 10.1. Bacteriophage.

These problems are solved in a variety of ways by different phage species. Phage particles also differ in their physical structures from species to species and often certain features of their life cycles are correlated with their structure. The most common type of nucleic acid in phages is double stranded linear DNA; however, double-stranded circular DNA, single-stranded linear and circular DNA, and single- and double-stranded linear RNA are also found. The molecular weight of the nucleic acid also varies over a hundredfold range from one species to the next, which contrasts with bacteria, for which the variation is rarely more than about ten percent. Phages with larger nucleic acid molecules have more complex life cycles and are less dependent on bacterial enzymes for their reproduction. An unusual feature of the DNA of some phages is the presence of bases other than the standard A, T, G, C. For example, T4 contains glucosylated 5-hydroxymethylcytosine instead of cytosine and SP01 has hydroxymethyluracil instead of thymine. The nucleic acid is a ways isolated from the environment, and thereby protected from harmful substances by an enclosing protein shell called either the *coat* or the *capsid*.

Stages in the Lytic Life Cycle of a Typical Phage

Phage life cycles fit into two distinct categories—the *lytic* and the *lysogenic* cycles. A phage in the lytic cycle converts an infected cell to a phage factory, and many phage progeny are produced. A phage capable only of lytic growth is called *virulent*. The lysogenic cycle, which has been observed only with phages containing double stranded DNA, is one in which no progeny particles are produced; the phage DNA usually becomes part of the bacterial chromosome. A phage capable of such a life cycle is called *temperate*. In this section only the lytic cycle is outlined. The lysogenic cycle will be described late.

There are many variations in the details of the life cycles of different virulent phages. There is, however, what may be called a basic lytic cycle, which is the following.

1. *Adsorption of the phage to specific receptors on the bacterial surface*. These receptors are very varied and serve the bacteria for purposes other than phage adsorption.
2. *Passage of the DNA from the phage through the bacterial cell wall*. Some types of tailed phages use an injection sequence shown schematically.
3. *Conversion of the infected bacterium to a phage-producing cell.* Following infection by most phages a bacterium loses the ability either to replicate or to transcribe its DNA; sometimes it loses both. This shutdown of host DNA or RNA synthesis is accomplished in many different ways depending on the phage species.

Production of Phage Nucleic Acid and Proteins

By several mechanisms the phage directs the synthesis of a replicative system that specifically makes copies of phage nucleic acid. This programming is accomplished either by synthesis of phage-specific polymerases or by addition of specificity elements to bacterial enzymes. Transcription is almost always initiated by the bacterial RNA polymerase but after the first transcription event either the bacterial polymerase is modified to recognize phage promoters or a phage-specific RNA polymerase is synthesized. Transcription is regulated and phage proteins are synthesized sequentially in time as they are needed. Distinct classes of mRNA are made at various times after infection; the major division is between "early" mRNA and "late" mRNA. Early mRNA usually encodes the enzymes required for takeover of the bacterium, for DNA replication, and for the synthesis of late mRNA. Late mRNA encodes the components of the phage particle, proteins needed for packaging of nucleic acid in the particles, and enzymes required to break open the bacterium. The more complex phage usually synthesize several classes of early mRNA, which are made in a particular time sequence.

Assembly of Phage Particles

This process is often called *morphogenesis*. Two types of proteins are needed for the assembly process; *structural proteins*, which are present in the phage particle, and *catalytic proteins*, which participate in the assembly process but do not become part of the phage particle. A subset of the latter class consists of the maturation proteins, which convert intracellular phage DNA to a form appropriate for packaging in the phage particle. Usually 50-1000 particles are produced, the number depending on the phage species.

Release of Newly Synthesized Phage

With most phages a phage protein called a *lysozyme* or an *endolysin* is synthesized late in the cycle of infection. This protein causes disruption of the cell wall. Other proteins called *membranases* dissolve the cell membrane and together with the lysozyme cause total destruction of the cell, so that phage are released to the surrounding medium. This disruption process is called *lysis*.

The events just described occur in an orderly sequence, as exemplified by the life cycle of *E. coli* phage T4 listed below (times in minutes at 37^{0}C):

t=0	Phage adsorbs to bacterial cell wall. Injection of phage DNA probably occurs within seconds of adsorption.
t=1	Synthesis of host DNA, RNA, and protein is totally turned off.
t=2	Synthesis of first mRNA begins.
t=3	Degradation of bacterial DNA begins.
t=5	Phage DNA synthesis is initiated.
t=9	Synthesis of "late" mRNA begins.
t=12	completed heads and tails appear.
t=15	First complete phage particle appears.
t=22	Lysis of bacteria; release of about 300 progeny phage.

The total duration of the cycle is typical, most phages having a life cycle of 20-60 minutes, which is comparable to the generation times of most bacteria. The life cycles of animal viruses are considerably longer—24 to 48 hours—but this is again comparable to the life cycle of an animal cell.

SPECIFIC PHAGES

On a biological scale of complexity, a phage is a relatively simple form of life. However, phages are sufficiently complex that there is not a single phage type for which all of the molecular details of its life cycle are completely understood. The major progress has been with a small number of phage species that grow in *E. coli*, *Bacillus subtilis*, and *Salmonella typhimurium*. Special features of each phage have resulted in particular phages being more suited to the study of certain processes. For example, regulation of transcription is best understood in phages λ and T7, the study of morphogenesis has been more successful with phages T4 and λ than with other phages, phages P1 and P22 gave the first clue to understanding transduction, T5 has yielded the best information about DNA injection, T4 has been most profitable in the study of DNA synthesis, and the study of T4 and T7 has shown clearly how a phage takes over a bacterium.

In the following sections several well-understood features of a few phages will be described.

E. coli phage T4 is a well-studied phage with an unusual DNA that lacks cytosine; instead, there is a modified form of cytosine that is called *5-hydroxymethylcytosine* (HMC), which base-pairs with guanine. This base is further modified by glucosylation—that is, sugar is coupled to the OH group of HMC. This has the consequence that purified T4 DNA is somewhat resistant to a variety of DNases.

HMC has an important function in the T4 life cycle. T4 is a rapidly multiplying phage, producing about 500 progeny per infected cell in 22 minutes. The high rate of T4 DNA synthesis in infected *E. coli*, which is greater than that in an uninfected cell, is accomplished in two ways: (1) the number of enzymes needed to make the DNA precursor nucleoside triphosphates is increased by synthesizing phage-specified enzymes and (2) *E. coli* DNA is degraded by phage-encoded nucleases to mononucleotides, which can be converted to triphosphates. The phages-specified DNases could not distinguish T4 DNA from *E. coli* DNA were it not for the presence of the HMC in the phage DNA—HMC-containing DNA is resistant to the enzymes.

E. coli does not possess enzymes for forming HMC. This synthesis is instead accomplished by two phage enzymes, which convert dCMP to dHDP.

$$\text{dCMP} \xrightarrow{\text{T4 hydroxymethylase}} \text{dHMP}$$
$$\text{dHMP} \xrightarrow{\text{HM kinase}} \text{dHDP}$$

The *E. coli* enzyme, nucleoside phosphate kinase, which forms all triphosphates in *E. coli*, then converts dHDP to dHTP, the immediate precursor of the HMC in the DNA.

When *E. coli* DNA is degraded, dCMP is produced. Whereas most of his is converted to dHMP, as shown above, some will be converted to dCTP, which could be incorporated into progeny T4 DNA. This would yield a T4 DNA molecule that could be attacked by the phage-encoded nucleases that degrade *E. coli* DNA. Thus, a mechanism exists for preventing such incorporation: a phage-specified enzyme, dCTPase, degrades any dCDP and dCTP that is formed, forming dCMP again. Ultimately, all of the dCMP is in the form of dHMP.

HMC creates another problem for T4, since *E. coli* possesses an endonuclease that attacks certain sequences of nucleotides containing HMC. To avoid this damage the HMC residues in T4 DNA are glucosylated. This is accomplished by two phage enzymes, α-glucosyl transferase (αgt) and β-glucosyl transferase (βgt), each of which transfers a glucose from uridine diphosphoglucose (UDPG) to HMC that is already in DNA. Thus glucosylation is a postreplicative modification. The *E. coli* endonuclease is inactive against glucosylated DNA; therefore, glucosylation is a protective device. A simple genetic experiment shows that this is the only essential function of glucosylation. A T4 αgt^- mutant cannot carry out glucosylation, so its newly synthesized DNA is destroyed by the *E. coli* HMC endonuclease. However, if an *E. coli* mutant ($rglB^-$), which lacks this nuclease, is used as a host bacterium, both T4 αgt^- mutants grow normally even though nonglucosylated DNA is produced.

Another property of T4 DNA is that it is terminally redundant that is, a sequence of bases (about one present of the total) is repeated at both ends of the molecule. Terminal redundancy is a feature of the DNA of many phage species and can be generated in several ways.

Another feature of T4 DNA is that although each phage particle contains one DNA molecule, the molecules differ from phage to phage, even if the population was produced by replication of a single phage particle and its progeny. A sample of T4 DNA molecules is *cyclically permuted*. The figure indicates schematically that the termini of the DNA molecules in the population can be found at many different bases within an overall sequence—in reality, probably at any point in the base sequence. Note that cyclic permutation is a property of a phage population, whereas terminal redundancy is a property of an individual

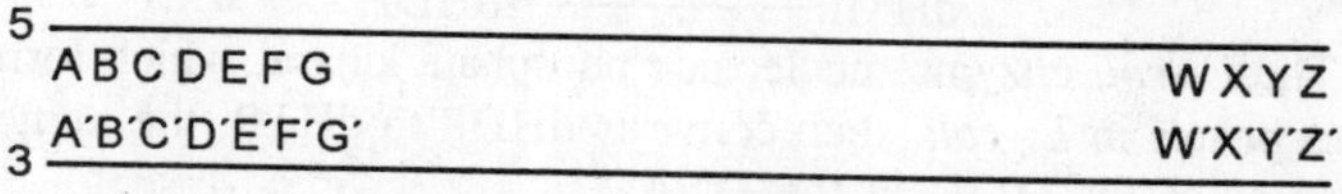

```
5 ____________________________________________________________
   ABCDEFG                                            WXYZ
   A'B'C'D'E'F'G'                                     W'X'Y'Z'
3 ____________________________________________________________
```

Fig. 10.2. A terminally redundant molecule.

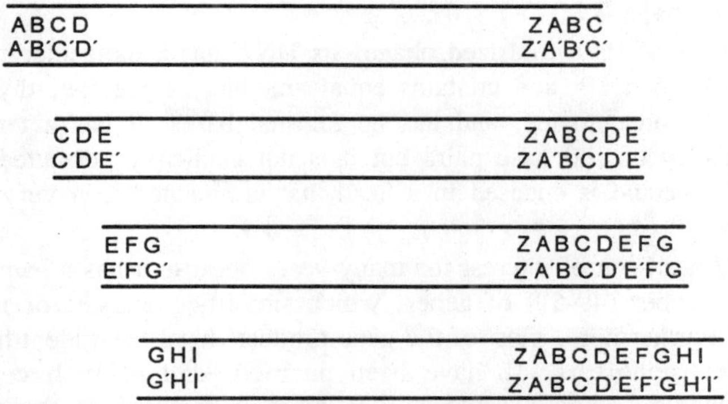

Fig. 10.3. A cyclically permuted collection of terminally redundant DNA molecules.

phage DNA molecule. With T4 both terminal redundancy and cyclic permutation are a consequence of the mechanism by which DNA is packaged into the phage head as next described.

The mechanism of packaging of T4 DNA in a phage head is not yet elucidated, though the process can be carried out in vitro. The basic problem is how the long strand of DNA is tightly folded so that it will fit in the phage head. It is thought that packaging begins by the attachment of one end of a DNA molecule to a protein contained in the phage head. Then either a condensation protein or a small, very basic molecule (such as a polyamine) induces folding. The best-understood feature of the process is that the molecule is cut from long concatemers. The cuts are not made in unique base sequences in the DNA because, if they were, T4 DNA could not be cyclically permuted. Instead, the cuts are made at positions that are determined by the amount of DNA that can fit in a head. Presumably a free end of the DNA molecule enters the head and this continues until there is no more room; then the concatemer is cut. This is known as the *headful mechanism* and it explains how both terminal redundancy and cyclic permutation arise. The essential point is that the DNA content of a T4 particle is greater than the length of DNA required to encode the T4 proteins. Thus, when cutting a headful form a concatemeric molecule, the final segment of packaged first—that is, the packaged DNA is terminally redundant. The first segment of the second DNA molecule that is packaged is not the same as the first segment of the first phage. Furthermore, since the second phage must also be terminally redundant, a third phage-DNA molecule must begin with still another segment. Thus, the collection of DNA molecules in the phage produced by a single infected bacterium is a cyclically permuted set.

E. coli Phage T7

Phage T7 is a midsized phage. Its DNA has a molecular weight of about 26 x 10^6 and contains equal amounts of adenine, thymine, guanine, and cytosine, and has no unusual bases. It has a terminal redundancy of 160 base pairs but it is not cyclically permuted. The DNA molecule is encased in a head that is attached to a very short tail.

T7 has been of interest for many years because it has a somewhat small number (49-50) of genes, which simplifies analysis of its life cycle. Furthermore, most of the gene products have been identified by gel electrophoresis, 42 have been purified, and many have been sequenced by chemical means. Of great importance is the recent elucidation of the complete sequence of 39, 930 base pairs of T7 DNA; this information allows one to locate, by direct inspection, all of the promoters, termination sites, regulatory sites, spacers, leaders, initiation codons, termination codons, and genes.

A great deal is known about the molecular biology of T7. However, we shall only discuss how it regulates its transcription, and this only in barest essentials. The life cycle of T7 can be divided into three transcriptional stage. Transcription is temporally regulated with transcript I made first. A basic principle of transcription of phage DNA is that the *first transcript must be made by the host RNA polymerase*, as no other enzyme is available; this is true of transcript I. One of the gene products in transcript I is a protein kinase that phosphorylates and thereby inactivates *E. coli* RNA polymerase; this is the first stage in takeover of the bacterium by T7, though the inactivation is not complete. A second protein made from transcript I is T7 RNA polymerase; this enzyme is essential for further transcription since it alone (not *E. coli* RNA polymerase) recognizes the promoters for transcripts II and III. T7 RNA polymerase does not recognize the termination site for transcript I and hence allows polymerization to proceed past this site. Transcript II encodes the replication proteins, the major protein from which the phage head is constructed, and a second inhibitor of *E. coli* RNA polymerase.

Synthesis of the latter protein completes the takeover of the bacterium and ensures that no energy is wasted making unnecessary bacterial RNA. A termination site for transcript II cause termination with 90 percent efficiency, meaning that 10 percent of the transcripts proceed on to the end of the DNA molecule. The extended region encodes the tail proteins. Since the tail is very short, less tail protein is needed than head protein; the extension of transcript II in 10 percent of the initiation events maintains the appropriate ratio of head and tail proteins. The lysis proteins are also included in the extended segment, but their concentration is not high enough to cause lysis during the period in which transcript II is being made. The DNA replication proteins are enzymes and hence are needed in very small amounts compared to the head and tail proteins. Therefore, late in the infection there is no need to make these enzymes; hence, at this time transcript III, which encodes all of the structural proteins and the lysis enzymes, but not the replication proteins, is made. Two proteins of T7 are responsible for the reduction in the synthesis of transcript II and the initiation of transcript III; synthesis of a general inhibitor of transcription, and slow injection of T7 DNA.

The promoter for transcript II (*pII*) is a very weak promoter compared to that for transcript III (*pIII*). One of the proteins encoded in transcript II causes an overall reduction in all transcription (the mechanism is unknown), when this proteins is made, transcription is initiated only from the strong promoter. However, since T7 RNA polymerase can act at both *pII* and *pIII*, a mechanism is needed to prevent activation of *pIII* in the interval when the DNA replication enzymes must be synthesized. The mechanism is one that has not been observed very often with phages. Most phage species inject their DNA within about 30 seconds of adsorption, whereas with T7 it takes about 10 minutes. Thus, *pIII* is not injected until about seven minutes after infection, which delays synthesis of transcript III.

Let us now summarize how the three transcripts are made in an orderly way. At first, *E. coli* RNA polymerase makes transcript I. An inhibitor, translated from transcript I, prevents further synthesis of transcript I. T7 RNA polymerase, translated from transcript I, enables transcript II to be made. An inhibitor, translated from transcript II, prevents further synthesis of this transcript. Slow injection delays entry of *pIII* into the cell; however, once *pIII* is available, transcript III is made. The net result is early synthesis of the DNA replication proteins and late synthesis of the structural proteins of the phage particle.

E. coli Phage ϕX174

Study of *E. coli* phage ϕX174 has been important for many reasons: it was the first organism to be discovered containing single-stranded DNA as its genetic material and having DNA in a circular form. Its DNA was also the first to be replicated *in vitro* to yield biologically active DNA, and study of its mode of replication led to the hypothesis of rolling-circle replication. The phage is one of the smallest known, contains 11 genes, and has a fairly simple life cycle. Each particle contains one circular, single-stranded DNA molecule consisting of 5386 nucleotides, whose base sequence is known. This strand is known as the (+) strand. The mode of DNA replication of ϕX174 DNA is clearly different from that of double-stranded DNA, and it is that which will concern us here.

The mechanism by which a base sequence is transmitted to daughter molecules in complementary base-pairing; thus, at some stage of the replication cycle of ϕX174 a DNA single strand (a) (–) strand must be synthesized whose sequence is complementary to the (+) strand. There are two ways that such a molecule could be produced: (1) a complementary strand could be synthesized and removed from the (+) strand as it is being made, like RNA transcribed from DNA, or (2) the circular single strand could be converted to a circular double-stranded molecule. Mechanism 2 is the one used by ϕX174, though some phages containing linear single-stranded DNA use the first mechanism.

Following infection of *E. coli* with ϕX174 and entry of the DNA in the cell, the first stage of replication occurs: conversion of the phage DNA molecule to a double-stranded supercoil by the *E. coli* replication system. Conversion occurs in four steps: synthesis of an RNA primer, extension of the primer by DNA polymerase III until replication of the circle is complete, removal of the RNA nucleotides and replacement with DNA nucleotides by polymerase I, and final sealing by DNA ligase. This supercoiled intermediate is known as RFI (replicating form I). In the second stage of replication RFI replicates many times to produce about 60 circular double-stranded molecules per cell. The second stage of ϕX174 replication—synthesis of (+) strands for packaging in the phage cost—then occurs.

A variant of the rolling circle mode, called *looped rolling circle replication*, is used to generate a progeny single-stranded circle from a double-stranded circular template. For phage ϕX174 this is accomplished in the following way. A phage protein called the gene-A protein makes a nick at the origin and in the process becomes covalently linked to the newly formed 5'-P terminus. Using the *E. coli* proteins—Rep, ssb, and polymerase III-chain growth occurs from the 3'-OH group, displacing the broken parental strand (which we call the (+) strand). This strand becomes coated with ssb protein and is not replicated. Leading strand synthesis continues until the origin is reached. At this point, the *A*-gene product binds to the 3'-OH group of the (+) strand and, using the energy obtained from the original nicking event, it joins the 3'-OH and 5'-P groups of the (+) strand, dissociates, and attaches to the newly synthesized (+) strand. This process can continue indefinitely, generating numerous circular (+) strands. Note that in looped rolling-circle replication, the displaced strand never exceeds the length of the circle, in contrast with ordinary σ replication. A summary of the replication cycle, with the timing of each stage. All transcription of ϕX174 DNA is from the (-) strand and hence does not occur until the first RFI is made. Therefore, synthesis of RFI must utilize host enzymes only, as is the case.

Table 10.1. Replication cycle of ϕX174

Stage	*Time, minutes at 33ºC*	*Event*
ss→RF	0-1	Adsorption and penetration; viral single strand converted to parental RF, RF transcribed
RF→RF	1-20	Parental RF replicates and 60 progeny RF are formed.
	25	Replication of RF molecules and host DNA stops
RF→ss	20-30	About 35 rolling circles form from progeny RF; from these, about 500 ss molecules are produced: DNA is inserted into phage particles
	40	Lysis occurs

E. coli Phage λ: the Lytic Cycle

E. coli phage λ may be the best understood of the double-stranded DNA phages, especially with respect to the elegant way in which transcription is regulated. The study of λ has made many important contributions to molecular biology, such as the discovery bidirectional replication, transcription termination and the Rho protein, anti-termination proteins, DNA ligase, DNA gyrase, site-specific recombination, and SOS repair, to name just a few.

λ is of particular interest because it can engage in both a lytic cycle, in which progeny phage are produced, and a lysogenic cycle, in which phage DNA becomes incorporated into the bacterial chromosome. Special features of the regulatory systems of λ, not found with all phages, determine the pathway that is used in any given infection. In this section, we will discuss significant features of the lytic lysogenic cycle is described in the next section.

The DNA of λ has two components: a double-strand containing 48,489 nucleotide pairs of known sequence a nucleotide single-strand extensions at the 5'-P terminus of. The two terminal single strands have complementary base and are called *cohesive ends*. Immediately following injected DNA the cohesive ends base-pair, yielding a circle adjacent 5'-P and 3'-OH groups are quickly sealed by DNA then the circle is supercircled.

As is the case for many phages, the genes are clustered at a function. For example, the head, tail, replication, and red genes form four distinct clusters. Many λ proteins—form regulatory proteins and those responsible for DNA synthesis particular sites in the DNA. In general these proteins are adjacent to their sites of action (when there is a single instance, the origin of DNA replication lies within the coding for gene O, which encodes a DNA replicating-initiation protein.

Transcription of λ occurs in several stages, with genes allowing synthesis of related genes at the appropriate time T7, λ makes an early mRNA, which encodes regulatory proteins and DNA replication enzymes, and a late mRNA, which encodes the structural proteins of the phage particle. In contrast with T7, *E. coli* RNA polymerase is used for all transcription events. With most phages (and those discussed so far) transcription is regulated by controlling the availability of particular promoters. For example, T7 early mRNA is made by *E. coli* RNA polymerase, which does not recognize the late promoters, and T7 RNA polymerase functions at the late, but not the early, promoter. Other phages—for example, T4—use the host polymerase throughout the life cycle but modify the enzyme by the addition of accessory molecules that enable it to recognize other promoters. Control of promoter availability is used in the lysogenic cycle of λ, but in the lytic cycle a totally different system is used. With λ all promoters are recognized by the host enzyme, but modification of the enzyme determines whether particular transcription-*termination* sites are recognized. That is, unmodified RNA polymerase initiates at a promoter and stops at a particular termination site, but the modified enzyme-passes the termination site and stops at a later site. This mechanism produces a temporal delay in the production of particular proteins in the following way. Consider a DNA segment containing the sequence *p*—A—B—t1—C—t2, in which *p* is a promoter, ABC are genes, and t1 and t2 are termination sites. Initially, RNA polymerase will attach to *p* and transcribe A and B. Let us assume that B encodes a regulatory protein that can bind to RNA polymerase so that t1 is ignored. In that case, before the C product can be made, the B product must be made and joined with RNA polymerase, thus, the C product will be synthesized at a later time than the A product.

Figure shows a portion of the genetic map of λ, with three regulatory genes (*cro*, N, Q), three promoters (*p*L, *p*R, *p*R2) and five termination sites (*t*L1, *t*R2, *t*R3, *t*R4). Seven mRNA molecules are also shown as arrows, the L and R mRNA series are transcribed leftward and rightward, respectively from complementary DNA strands. The arrows indicate the site of action of the regulatory protein. The critical genes whose products modify RNA polymerase are N and Q; their products are called *antiterminators*. Synthesis of each of the mRNAs—L1, R1, and R4—begins at the same time. The terminator tR1 is a weak one, so very little of the O and P products are made Once the N product is translated from L1, it binds to the *nut*L and *nut*R sites on the DNA molecule.

During the next passage of RNA polymerase along the DNA, the enzyme picks up the N protein from these sites and is thereby modified. The terminators *t*L1, *t*R1 and *t*R2 are then bypassed, leading to production of L2 (which we will not discuss further) and R3. Translation of R3 yields the DNA replication proteins, allowing replication to begin, and the product of gene Q. The Q protein then binds to the *site qut*, where it is picked up by RNA polymerase, which is thereby modified so that it ignores *t*R4. Extension of R4 yields R5, the late mRNA, from which heads, tails, other structural proteins, and the lysis enzyme are made. Note that the delay of production of the late proteins compared to the early proteins is simply a result of the time required to transcribe and translate the N and Q genes. Details of the

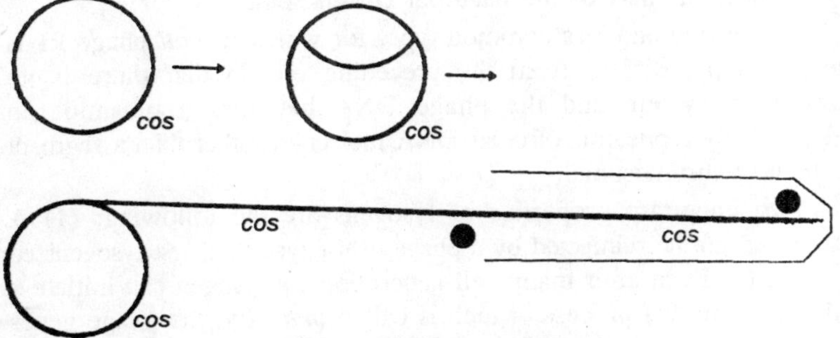

Fig. 10.4. The replication sequence of phage λ.

regulation of transcription and also of the roles of the genes *cro* and *c*I can be found in MB.

It was mentioned earlier that λ DNA circularizes shortly after infection and that the cohesive ends base-pair to form the double stranded *cos* region. Clearly, at some stage of the life cycle the single-stranded termini must be regenerated, since the DNA is linear within a phage head. This is accomplished by a cutting system, called *Terminase* (or *Ter*), which cuts within the *cos* region of progeny DNA. However, the cuts are not made in progeny circles; instead, DNA replication and cutting are coordinated in a way that is found with many phage species. The earliest stage of replication of λ DNA is the bidirectional θ mode. Progeny circles do not continue to replicate in this way but switch to the rolling circle mode, yielding a long linear branch. Note that the branch contains several *cos* sites compared to a circle, which contains only one. Progeny λ DNA molecules are cut from the branch of the rolling circles by terminase. The circular portion of the rolling circle is not cut (this would prevent further replication), because terminase requires two *cos* sites or one *cos* site plus one free single-stranded terminus. Thus, excision of λ units occurs by sequential cleavage from the free end of the branch.

E. coli Phage λ: the Lysogenic Life Cycle

There are two types of lysogenic cycles. The most common one, for which *E. coli* phage λ is the prototype, follows:

1. A DNA molecule is injected into a bacterium.
2. After a brief period of transcription, needed to synthesize an integration enzyme, transcription is turned off by a repressor.
3. A phage DNA molecule is inserted into the DNA of the bacterium, forming a *prophage*.
4. The bacterium continues to grow and multiply and the phage genes replicate as part of the bacterial chromosome.

The second and less common type, for which *E. coli* phage P1 is the prototype, differs from the preceding one in that there is no integration system and the phage DNA becomes a plasmid (an independently replicating circular DNA molecule) rather than a segment of the host chromosome.

Two important properties of lysogens are the following: (1) A lysogen cannot be reinfected by a phage of the type that first lysogenized the cell. (2) Even after many cell generations, a lysogen can initiate a lytic cycle, in this process, which is called *inducion*, the phage genes are excised as a single segment of DNA.

The molecular mechanism for immunity and the circumstances that give rise to induction will be described shortly.

The resistance of a λ lysogen to infection by λ is called *immunity*. The cause of the phenomenon is the following. Phage λ contains a repressor-operater system. The repressor gene is called *cI*; the repressor protein binds to two operators *o*L and *o*R, which are adjacent to two promoters *p*L and *p*R. The letter L and R means leftward and rightward and refer to the direction of synthesis of two early mRNA molecules, when the genetic map is drawn in a standard orientation.

In a lysogen the cI repressor is synthesized continuously and in slight excess with respect to the operators. Both *o*L and *o*R sites contain bound repressor molecules so that *p*L *p*R are unavailable to RNA polymerase. Thus, in a lysogen, transcription from the two early promoters is prevented. It will be seen later that this is sufficient to keep the prophage in an "off" state thus the lysogen grows indefinitely. If a normal cell is infected by λ, the two operators of the incoming λ DNA molecule will be unoccupied, since phage repressor has not yet been made and transcription will occur. However, if a phage tries to infect a lysogen, the excess repressor molecules already present in the lysogen bind to the two operators on the infecting DNA molecule before as RNA polymerase can bind to the *p*L and *p*R sites. This operator-binding prevents to phage from proceeding into lytic development. This inhibition is referred to as *resistance to homoimmune superinfection*.

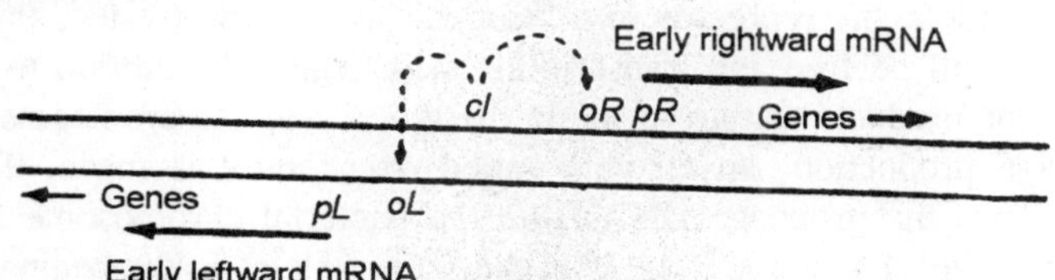

Fig. 10.5. The repressor-operator system of λ, showing the two early mRNA molecules. Symbole cl, repressor gene; p, promoter, o, operator, L. left, R. right.

What happens to superinfecting DNA? It is able to form a supercoil—phage gene products are not needed for supercoiling—but it cannot replicate. The bacterium is unaffected by the presence of this DNA molecule and grows and divides normally, so that the superinfecting DNA is progressively diluted out.

There are many temperate phages of *E. coli* other than λ. Two related to λ are phages 21 and 434. Each of these phages has its own immune system—that is, its own repressor and repressor-specific operators. Thus, the 434 repressor cannot bind to a λ operator and a λ repressor cannot bind to a 434 operator. Such a pair of phages is said to be *heteroimmune* with respect to one another. A temperate phage can form a plaque on a heteroimmune lysogen because the repressor made in the lysogen does not bind to the operator of the superinfecting phage. The immunity region of the DNA, which includes the *cI* gene, operators, and promoters, is denoted by *imm*—specifically, *imm*λ, *imm*21 and *imm*434. Interesting hybrid phages, which have been very useful in the laboratory, have been created by crossing two heteroimmune phages and selecting a recombinant containing the immunity region of one phage and the remaining genes of the other heteroimmune phage. A prominent example is a λ-434 hybrid, which is genotypically designated λ*imm*434.

Table 10.2. Ability of different phages to form plaques on homoimmune and heteroimmune lysogens

Superinfecting phage	*Lysogen*		
	B(λ)	*B(434)*	*B(21)*
λ	–	+	+
434	+	–	+
21	+	+	–

Induction is also based on the repressor-operator interaction. If, for any reason, the repressor in a lysogenic cell is inactivated, the λ operators will be free and transcription will begin. In addition to all of the gene products that are made in the lytic cycle and are necessary for phage production, an enzyme called *axcisionase* is made. This enzyme cuts the prophage DNA from the bacterial chromosome and forms a circle; thus the λ DNA is in the same state as in the beginning of a lytic cycle. The molecular mechanism of induction is not well understood. It is known that the repressor is cleaved by a protease, but the events leading to induction are unclear. The initial signal seems to be damage to DNA, in particular, single-stranded DNA produced by the repair of damage. If DNA is damaged, cell death might occur, so it is to the advantage of the prophage to get out of the bacterium. Precisely how the damaged DNA activates the protease that cleaves the repressor is not known, but is an active field of study. Ultraviolet radiation is a powerful inducing agent because of its ability to damage DNA.

In the lysogenic cycle of λ and other phages, a phage DNA molecule is inserted (or integrated, to an equivalent term) into the bacterial chromosome to form a prophage. In order to form lysogen, (1) repression must be established and (2) a prophage must be formed by insertion of a λ DNA molecule.

When λ DNA integrates, it is inserted at a preferred position in the *E. coli* chromosome, this site is between the *gal* operon and the *bio* (biotin) operon and is called the λ attachment site or, genotypically, *att*. For most temperate phages integration occurs at one preferred site, though there are a few phages that either have several preferred sites or appear to be able to insert their DNA anywhere in the *E. coli* chromosome. An important feature of the insertion mechanism was derived from the genetic observation that the gene order in the λ prophage is a permutation of the gene order of the DNA in the phage. This permutation was explained by a model (due to Allen Campbell and widely known as the *campbel model*) for the mechanism of integration. In this correct model, λ DNA is circularized first; and then prophage integration occurs as a physical breakage and rejoining of phage and host DNA—precisely, between the bacterial DNA attachment site and another attachment site in the phage DNA is

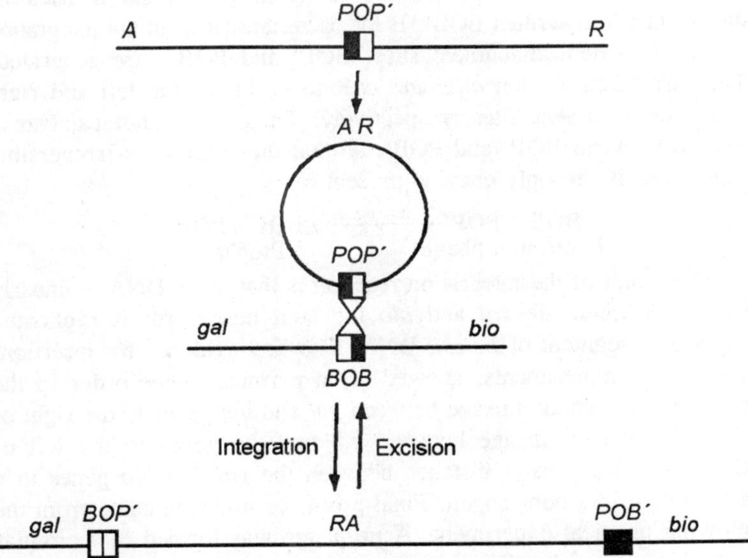

Fig. 10.6. The Campell model for the mechanism of prophage integration and excision of phage λ.

located near the center of the phage DNA molecule. A phage protein, *integrase* (its gene designation is *int*), recognizes the phage DNA and bacterial DNA attachment sites and catalyzes the physical exchange. This results in integration of the λ DNA molecule in into the bacterial DNA. As a consequence of the circularization and integration, the linear order of the genes in the infecting λ DNA molecule is permuted. A question that arose after this model was proposed was: why doesn't the integrase excise the prophage shortly after integration occurs, since the prophage is flanked by two attachment sites? Furthermore, excision would seem to be the preferred reaction since, kinetically, two attachment sites in the same DNA molecule (that is, the host chromosome) ought to interact with each other more rapidly than two attachment sites in different DNA molecules (phage and bacterial DNA). Why this does not happen became clear when it was discovered that, inasmuch as the DNA attachment sites in the phage and the bacterium are not the same, they recombine to form prophage attachment sites that, as a consequence, also differ from the sites joining the bacterial and phage DNA.

All of the λ attachment sites have three different components. One of these is common to all sites and is denoted by the letter O. The phage attachment site is written POP' (P for phage) and the bacterial attachment site is written BOB' (B for bacteria). Thus, in the integration reaction two new attachment sites, BOP' and POB', are generated. There are often written *att*L and *att*R to designate the left and right prophage-attachment sites, respectively. Integrase cannot catalyze a reaction between BOP' and POB', so that the reaction is irreversible if integrase is the only enzyme present.

$$\underset{\text{Bacterium phage}}{\text{BOB} + \text{POP}'} \xrightarrow{\text{int egrase}} \underset{\text{Prophage}}{\text{BOP}' + \text{POB}'}$$

The result of the integration reaction is that the λ DNA is linearly inserted between the *gal* and *bio* loci and henceforth is replicated simply as a segment of *E. coli* DNA. The first evidence for insertion, gained from experiments, showed (1) a permuted gene order in the prophage, (2) genetic linkage between *gal* and the genes to the right of POP' (3) genetic linkage between *bio* and the genes to the left of POP' and (4) a greater distance between the *gal* and *bio* genes in a lysogen than in a nonlysogen. Final proof for insertion came from the following physical experiment. A prophage was formed in a circular plasmid whose molecular weight was 105 x 10^6 and which contained BOB'. The plasmid remained circular but its molecular weight increased by the amount of one λ molecule (31 x 10^6) to 136 x 10^6. Integrase has been purified, the base sequences of the BOB' and POP' sites are known and the integration reaction can be carried out *in vitro*. Its mechanism is fairly understood.

The synthesis of integrase is coupled to the synthesis of the *cI* repressor. This is efficient because integrase and the repressor are both needed in the lysogenic cycle and neither is needed in the lytic cycle. In the absence of a positive regulatory element (the product of the λ gene *cII*) the promoters for both the *int* and *cI* genes are unavailable to RNA polymerase. Shortly after infection, the *cII* gene product is translated from R3 mRNA. If the concentration of the protein is high enough, the *cII* product binds to sites near the promoters for the *cI* and *int* genes (designated *pre* and *cI*, respectively) and thereby renders them accessible to RNA polymerase. (In this respect the *cII* protein is similar to the cAMP-CAP complex in the *lac* operon). RNA polymerase then transcribes both genes and the gene products are made. Synthesis of the *cII* protein is also the key step in the decision between a lytic and lysogenic pathway in a particular infection; however, details of its role are not yet understood.

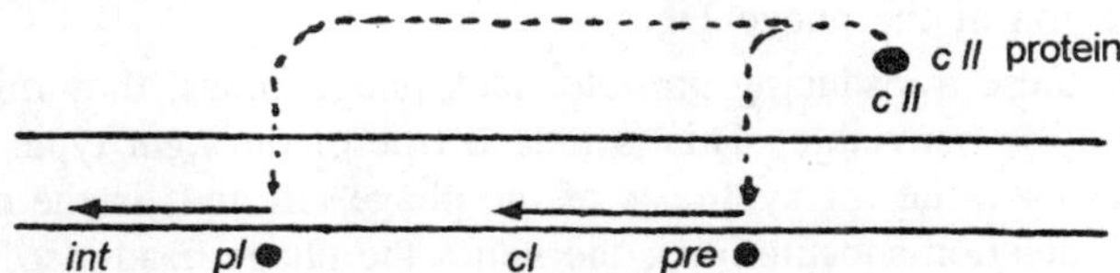

Fig. 10.7. The role of the cII protein in simulating synthesis of cI and int mRNA from pre and pl respectively.

Transducing Phages

Some phase species, known as *transducing phases*, are able to package bacterial DNA as well as phage DNA. Particles containing bacterial DNA, called *transducing particles*, are not always formed and usually constitute only a small fraction of a phage population. Such particles can arise by two mechanisms, aberrant excision of a prophage and packaging of bacterial DNA fragments. We begin with a discussion of the first mechanism. One mechanism by which transducing particles form depicts the formation of the galactose and biotintransducing forms of phage λ namely, λ *gal* λ *bio*.

When λ prophage is induced, an orderly sequence of events ensues in which the prophage DNA is precisely excised from the host DNA. In the case of phage λ this is accomplished by the combined efforts of the *int* and *xis* genes acting on the left and right prophage attachment sites. At a very low frequency—namely, in about one cell per 10^6—10^{-} cells—an excision error is made: two incorrect cuts are made—one within the prophage and the other cut in the bacterial DNA. The pair of abnormal cuts will not always yield a length of DNA that can fit in a λ phage head—it may be too large or too small. However, it the spacing between the cuts produces a molecule between 79 percent and 106 percent of the length of a normal λ phage-DNA molecule, packaging can occur. Since the prophage is located between the *E. coli gal* and *bio* genes and because the cut in the host DNA can be either to the right or the left of the prophage, transducing particles can arise carrying the *bio* genes (cut to the right) or the *gal* genes (cut to the left). Formation of the λ*gal* and λ*bio* transducing particles entails less of λ genes. The λ*gal* particle lacks the tail genes, which are located at the right end of the prophage; the λ*bio* particle lacks the *int xis*,... genes from the left end of the prophage. The number of missing phage genes of course depends on the position of the cuts that generated the particle and thus correlates with the amount of bacterial DNA in the particle. The missing phages genes come from the prophage ends, but because of the permutation of the gene order in the prophage and the phage particle, the deleted phage genes are always form the central region of the phage DNA.

Since these transducing particles lack phage genes, they might be expected to be nonviable. This is indeed true of the λ*gal* type, which lack genes essential for synthesis of the phage tail and, in the case of the larger deletion-substitutions, the genes the phage head also. Thus, these particles are incapable of producing progeny; they are defective and this is denoted by the symbol d—thus, a *gal*-transducing particle is written λd*gal* (and sometimes λdg). The λd*gal* particle contains the cohesive ends and all of the information for DNA replication and transcription; it therefore goes through a normal life cycle, including bacterial lysis. In fact, if the head genes are not deleted, the concatemeric branch produced by rolling circle replication is cleaved by the Ter system. However, tails are not added to the filled head, no viable particles are produced in the lysate, and plaques are not formed on a host lawn. Note that there is no discrepancy between the formation of a λd*gal* transducing particle and its ability to reproduce itself. A λd*gal* particle fails to reproduce only because it lacks the tail genes, but such a particle arises from a normal prophage having a full set of genes.

The situation is quite different with λ*bio* particles for these usually lack only nonessential genes—*int*, *xis* etc. These genes are needed for the lysogenic cycle but not for the lytic cycle, so these particles are able to replicate and to form plaques. To denote thus, the letter p, for plaque-forming, is added; this is, the particle is called λphio.

Transducing phages, like λ, able to form transducing particles containing bacterial genes from selected regions of the bacterial chromosome, are called *specialized transducing phage*. Other phage species are able to package fragments of bacterial DNA from all parts of the chromosomes. These are called *generalized transducing phages*. The specialized transducing phages producing transducing particles only come from a lysogen. In contrast, the generalized transducing phages form transducing particles from infected cells. The mechanism by which these particles are produced is straightforward. Some phage species, of which T4 is an example, degrade bacterial DNA early in infection. With T4 the degradation is complete, but in other species, of which *E. coli* phage P1 and *Salmonella typhinurium* phage P22 are examples, fragments of chromosomal DNA comparable in size to phage DNA are present at the time phage DNA is packaged. The packaging system of λ is a site-specific one, requiring *cos* sites.

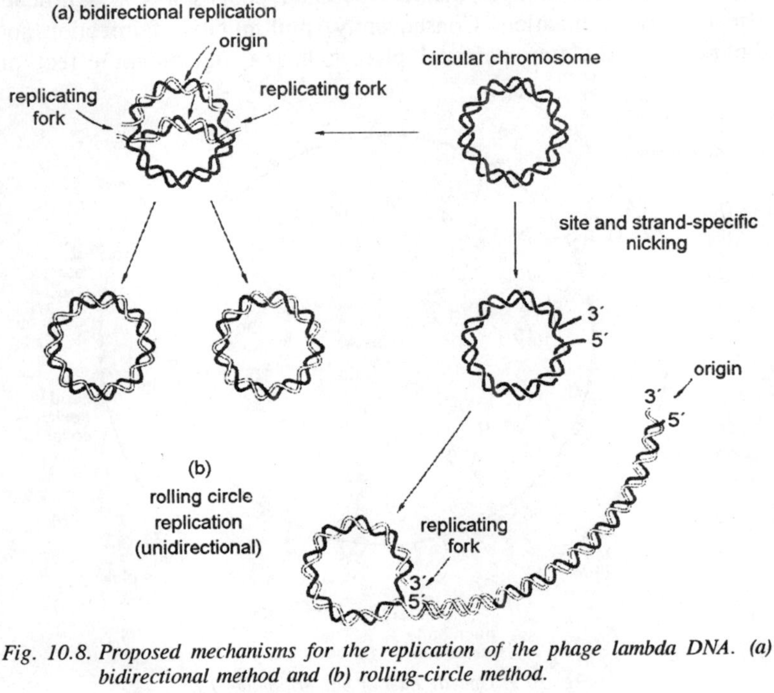

Fig. 10.8. Proposed mechanisms for the replication of the phage lambda DNA. (a) bidirectional method and (b) rolling-circle method.

However, some phages, like T4, either merely fill a head, or like P1 and P22, are able to package fragments having a range of sizes. The latter phages cannot distinguish bacterial DNA from phage DNA and occasionally package a piece of bacterial DNA. In a P1 population about one percent of the particles contain only a fragment of bacterial DNA.

Transducing particles are quite useful in genetic experiments. For example, they can be used to transfer bacterial as means of isolating particular segments of a bacterial chromosome.

Genetic Maps in the Bacteriophages

Early maps were constructed by standard complementation and recombination techniques using simple plaque morphology mutants. A larger range of mutants was obtained when conditional lethal mutants were used, e.g., temperature-sensitive mutants and amber mutants. It should be realized that recombination in phages is effectively as population phenomenon, and repeated rounds of replication are accompanied by rounds of recombination. Consequently, multiplicity of infection and the ratio of the two parental phages has a significant effect on recombination. Consequently, multiplicity of infection and the ratio of the two parental phages has a significant effect on

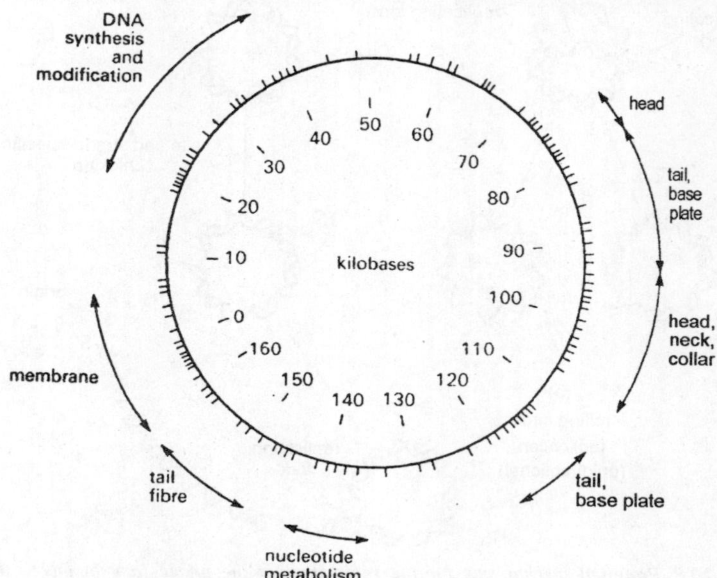

Fig. 10.9. Genetic map of bacteriophage T4.

recombination values obtained. A theory for phage recombination was proposed by Visconti and Delbruck (1953) which is discussed in detail by Lewin (1977). Recombination data following mixed infection of bacterias built up linkage maps which for many phages were circular, e.g., T4, although some maps were linear, e.g., λ. It must not be assumed that, because the genetic map is circular, the physical map is also circular. T4 molecules have been shown to be linear in phage heads, and the linear molecules in the heads of λ can be made to circularize in the test-tube. The maps of the small bacteriophages of ϕX174 (9 genes), M13 (8 genes) and MS2 (4 genes) provide fascinating insights into the minimum number of genes required to maintain an infectious virus particle. MS2 is particularly interesting as it has a single stranded RNA molecule which acts both as a mRNA and as a template for replication. Recombination does not occur and maps have been constructed using other techniques.

Genetic Analysis of T4

Detailed maps have been built up using complementation, recombination analysis, heteroduplex analysis and deletion data. The most detailed maps were constructed using a amber mutations, as these affected a wide variety of proteins. The proteins of T4 can be broadly divided into two groups: the early proteins and the late proteins. The early proteins are those involved in eliminating bacterial functions and replicating phage DNA. An impression of often given that viruses can 'take over' the biosynthetic machinery of the host cell and exploit this for their own replication. In T4, however, at least 24 genes are involved in early functions which range from a phage-coded DNA polymerase to enzymes involved in synthesizing precursors for DNA replication. T4 contains 5-hydroxymethyl cytosine in place of cytosine in its DNA, and an enzyme is synthesized by the phage to destroy dCTP; the phage also synthesizes dCMP hydroxymethylase. Chain terminating triplets UAG, UAA and UGA have been exten-sively exploited in the analysis of bacteriophage T4 and other systems. The UAG triplet has been called an *amber* mutant after Bernstein who discovered it, and by extension UAA mutants are known as *ochre* mutants. Their use depends on host bacterial strains which can suppress the effects of the chain termination.

In non-suppressor strains chain termination occurs to give a shortened form of, for example, the head protein subunit. Lysates of the infected bacteria can be seen under the electron microscope to have all the phage components except the phage heads. The normal phage head protein has a serine residue at the site of the amber mutation. In the suppressor strain C there is a mutant tyrosine transfer RNA which has an anticodon CUA instead of the normal GUA. The mutant tRNA can recognize UAG as a codon and tyrosine is inserted at the serine position to give a functional head protein. The use of these mutants has increased the range of mutants which can be studied as in theory any protein, even essential proteins, could be affected. They have also been useful as a safety device in gene cloning vectors to reduce the likelihood of escape of recombinant DNA. Suppressio strains are rare in nature and grow rather poorly, therefore the probability of a recombinant amber vector escaping from the laboratory and finding such a strain is low. Amber mutations have been extensively used to analyze late proteins. Growth of on amber mutation on a non-suppressor strain often resulted in the production of incomplete phage particles which had tails but no heads, or heads but no tails. Analysis of the details of the process has enabled a morphogenetic sequence to be constructed. More than 20 genes have been implicated in the assembly of the head, although only gene 23 actually codes for the major head protein. More than 30 genes are involved in tail assembly, so that this phage has a complex organization. The structure of the head has been implicated in packaging DNA. The basic idea is that the volume of the head dictates the length of the chromosome packaged. This is called the 'headful hypothesis' and it has important genetic implications. Polyhead mutations result in heads which are up to 20 times longer than a normal head, and these heads contain much larger DNA molecules than wild-type phages.

Table 10.3. A selection of enzymes and proteins encoded in the T4 genome.

Gene	*Enzyme or protein*	*Function*
1	DNA-kinase	DNA synthesis
30	Ligase	Joining single-strand breaks in DNA
32	DNA unwinding protein	DNA synthesis
42	dCMP hydroxymethylase	Production of hydroxymethyl cytosine
43	DNA polymerase exonuclease	DNA synthesis
46	Nuclease	Host DNA degradation
56	dCTP and dCDP ases	Degradation of cytosine compounds
e	Lysozyme (endolysin)	Destruction of cell wall

Chromosome Structure of T4

The structure of the T4 chromosome caused considerable confusion for a number of years. The genetic map was clearly circular, and yet the molecules from phages were linear. The first clue towards solving this paradox was the occurrence of heterozygosity at a frequency of about 2% per market. A cross between an r^+ and an *rII* mutant gave about 2% mottled plaques which had properties of both plaque morphologies. Eventually these plaques were shown to derive from two types of phage. One type was a genuine heteroduplex r^+/r, the hybrid DNA discussed earlier, which segregated on DNA replication to give r^+/r^+ and r/r duplexes. The second type was shown to have two sets of the same genes, one at each end of the chromosome. This was described as ***terminal redundancy***. The genes present in double dose varied from phage to phage, but each gene would be a duplicate in 2% of the phages. The origin of this terminal redundancy was obscure until it was realized that the chromosome were cut out from a long molecule consisting of repeating unit genomes called *concatenates*.

```
unit genome
1 2 3 4 5 6 1 2  3 4 5 6 1 2 3 4  5 6
chromosome 1        chromosome 2
```

The chromosome is longer than the unit genome, and consequently each molecule will contain different duplicated segments. Each molecule is a circular permutation of the gene order. The length of the chromosome is dictated by the volume of the head, and evidence for this is that a deletion of genes from the chromosome increases the length of the terminal redundancy. Concatenates can be generated either by a rolling-circle DNA replication mechanism or by recombination. In T4 they are generated by recombination between two terminally redundant molecules:

```
1 2 3 4 5 6 1 2                     1 2 3 4 5 6 1 2 3 4 5 6 1 2
              ×                       →
              1 2 3 4 5 6 1 2
```

Direct proof of these proposals for chromosome structure has been obtained by examination of DNA molecules in the electron microscope. Two methods were used; firstly, phage DNA molecules were denatured to give single strands and were then allowed to reanneal. A variety of circular molecules resulted, depending on which single strands

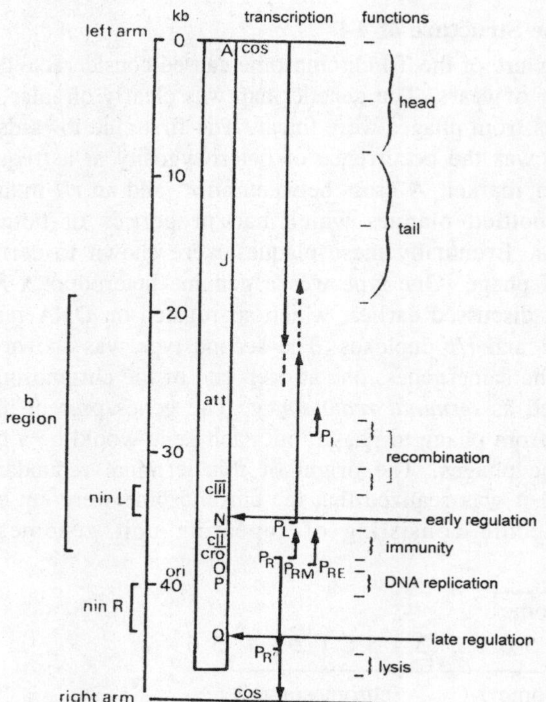

Fig. 10.10. Genetic map of bacteriophage lambda.

reannealed. For molecules with different terminally redundant sequences, only one sequence will be able to find a partner, leaving two single strands unpaired. This permits measurements to be made of the unit genome, the circle, and the terminally redundant regions. For T4, Kim and Davidson (1974) have estimated that the unit genome is 166 $\pm$ 2 kb = 1000 nucleotide pairs, and the total genome is 169 $\pm$ 3 kb, giving a terminal redundancy of 3 kb. A second method, using partial degradation of intact molecules by exonuclease III, also gave rise to circular molecules but without the single-strand tails.

Genetic Analysis of Phage Lambda, λ

Phage λ has a chromosome which is about one-quarter of the length of the T4 chromosome. It has been estimated that there are at least 35 genes coded by 46.5 kb. 20 genes have been implicated in the production of the head and tail proteins of the phage particle, 5 genes control replication, 6 genes are involved in regulation, 2 with DNA replication and 2 with lysis of the host. The location of these genes on the map has been made by recombination, deletion and heteroduplex analysis.

The genetic map is highly structured with genes for head formation, tail formation, recombination, regulation, immunity and lysis all being to a large extent grouped in blocks along the chromosome. The genetic map is also linear, and a combination of biochemical, biophysical and genetic techniques has shown that the chromosome is a linear duplex which has single-stranded regions of 12 bases at each end. These single-stranded regions can pair with each other to give rise to a circular molecule which is finally sealed by a ligase. The lambda chromosome has a number of unique properties which have allowed it to be manipulated extensively *in vivo* and *in vitro*. When the molecule is

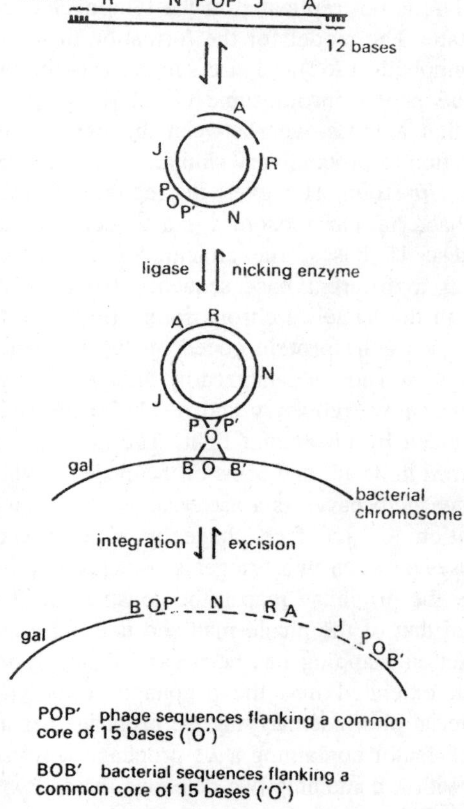

Fig. 10.11. Model for the circularization of the lambda chromosome and its insertion into the bacterial chromosome.

stirred vigorously, it can be broken into two halves which can be separated by their different buoyant densities. The leaf half has 55% G-C pairs and has a greater density than the right half, which has 45% G-C content. Mapping of the genes on these fragments can be made using a marker rescue technique. Helper phages with mutant genes allow the free wild-type DNA to infect the host cell. Wild-type plaques are obtained when the free DNA carries the wild-type allele corresponding to the mutant gene in the helper phage. The single strands of DNA can also be separated by centrifugation into the left strand (*l*) which is transcribed to the left, and the right strand (*r*) which is transcribed to the right.

Integration of the λ Chromosome

When λ DNA is not replicated in the lytic cycle, it is reduced to the prophage state. The model for the formation of the prophage was proposed by Campbell (1962). Figure summarizes the process, which assumes that the phage chromosome circularizes after entering the host cell, and that a cross-over between the two circular molecules results in integration, a process very similar in principle to the insertion of *F* to produce *Hfr* strain. However, the region of homology between the phage and bacterial chromosomes is a sequence of only 15 bases. Apart from these 15 bases, the attachment site *att* on the phage chromosome has a different base sequence from the corresponding attachment site in the bacterial chromosome. Integration of the phage chromosome requires the protein coded by the *int* gene whereas the reverse process of excision is catalyzed by the *int* + *xis* gene products. Excision occurs at low frequency and results in the lytic cycle. This can also be induced by ultraviolet light. The accuracy of this model has been confirmed in detail, and some of the evidence will be mentioned briefly. The prophage behaves as a bacterial gene and can be transferred during conjugation to give free phage by a process called *zygotic induction*. Crosses between two lysogenic bacteria with different phage mutations allow the prophage map to be constructed. The gene order is different from that of the phage map and is exactly as predicted by the model. Deletion mapping has been carried out by using bacterial deletions which extended into the prophage. Finally, heteroduplex analysis of bacterial plasmids carrying the prophage has also confirmed the model. An F-factor containing a λ^+ prophage was used to produce a heteroduplex with λ containing two deletions, *b*2 and *b*5. The position of the deletions relative to the insertion points showed that the prophage markers were a permutation of the linear phage genome.

Origin of Transducing Phages

Phage λ can transfer only a few bacterial genes by specialized transduction. These are the genes located close to the attachment site, *Gal*$^+$ markers are found in the phage chromosomes following inaccurate recombination which occurs during the induction of the lytic cycle following exposure of $(\lambda)^+$ cells to ultraviolet radiation. The *gal*$^+$ marker can be transduced at a frequency of one in a million, a process known as *low-frequency transduction*. The *gal*$^+$ colonies obtained are in fact heterozygous and genotypically are *gal*$^-$/λ-*gal*$^+$. When these bacteria re induced with ultraviolet radiation, phage stocks are produced which transduce at a much higher frequency (*high-frequency transduction*). When these lysates are used to infect cells at low multiplicities of infection, only one phage infects a bacterium and some *gal*$^+$ recombinants are immune to infection by further phage but cannot be induced to lyse with ultraviolet light. This can be explained by assuming that the incorporation of the *gal*$^+$ genes has resulted in the loss of essential gene functions necessary for the lytic cycle of the phage. These phages are described as λ*dgal* or λ*dg* for defective *gal* transducing phages. These phages very much resemble *F'* prime factors.

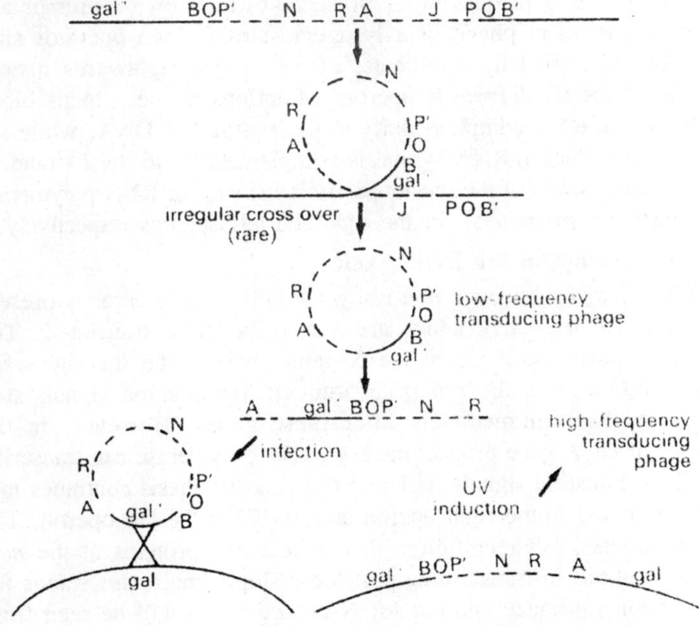

Fig. 10.12. Mechanism of transduction of gal$^+$ genes by lambda.

Genetic Basis of Lysogeny

The normal plaques of lambda are turbid due to the establishment of lysogenic bacteria which are immune to further phage infection. Mutant phages can be detected which have *clear plaques*. These have been used to analyze the genetic control of lysogeny. When the clear-plaque mutants were tested on lysogenic and non-lysogenic strains, they were found to fall into two groups. *c* mutants produced clear plaques on the non-lysogenic strain, but failed to lyse the lysogenic strain. The second-class *v* mutants produced clear plaques on both strains. The explanation for the *c* mutants is that they have a mutation affecting the synthesis of a repressor coded by the phage but with an unaltered operator site. The absence of the repressor means that *c* mutants cannot switch themselves off, and consequently behave like virulent phages. The operator site, however, remains sensitive to the repressor, and hence *c* mutants fail to infect lysogenic strains which contain a repressor synthesized by the resident phage. *cI* codes for the repressor, and *cII* and *cIII*, distinguished by complementation analysis, are necessary for the expression of *cI*. The *v* mutants have an altered operator site which is no longer sensitive to the repressor. Consequently they produce clear plaques in the presence of their own repressor and that of the resident phage in a lysogenic strain. Two operator sites have been identified by *v* mutants, *v*1 *v*3 for the rightwards operon and *sex v*2 for the leftwards operon. Mutations at the *x* locus block synthesis of mRNA complementary to the *r* strand of DNA, while *sex v*2 mutations block mRNA synthesis complementary to the *l* strand. It has been suggested that the mutations affect the sites of RNA polymerase attachment, the *promoters*, for the right and left operons respectively.

Gene Expression in the Lytic Cycle

When lambda infects a non-lysogenic cell, no repressor is present and therefore the early genes are transcribed and translated. The leftwards operon ($O_L P_L$) give the *N* gene product and the rightward operon ($P_R O_R$) gives the *cro* gene product. Termination signals stop mRNA synthesis immediately after these genes. However, in the presence of the *N* gene product the host RNA polymerase can transcribe past the termination signals, tL1 and tR1, and synthesis continues into *cII*, O, P and Q in the right operon and to *cIII* in the left operon. The N-protein acts in cooperation with the host *nus* proteins at the nut_R and nut_L regions in transcribing past these stop signals, *nus* stands for N-utilization substance and *nut* for N-utilization. It will be seen from figure that, although *t* must be downstream for *nut*, there can be a variable distance between the two regions. The N-protein is short-lived and therefore needs to be synthesized continuously to exert its effect. The fate of the cell then depends on the balance of control of other gene products in the lytic or lysogenic cycles. The Q gene product is also an anti-terminator which acts at the *qut* region in much the same way as the N-protein acts at *nut* regions. In the absence of the protein a short transcript is produced from P'_R due to the presence of a termination region but, when the Q-product is present, transcription occurs through this region to give expression of the late proteins in a cascade fashion.

This results in the synthesis of the tail and head proteins. However, the *cII* and *cIII* products have also been synthesized and these can stimulate the synthesis of the *cI* repressor, resulting in a complete shutdown of both operons. The gene product for *cro*, however, can reduce the synthesis of *cII* and *cIII* activities, thereby preventing *cI* expression. Consequently, the decision between the lytic and the lysogenic pathways depends on the balance between the amount of *cro* and N gene products. Excess *cro* product will prevent repressor synthesis giving the lytic cycle while N synthesis followed by *cII* and *cIII* synthesis will result in repressor synthesis and lysogeny. The *cII* product interacts with the DNA at *Pre* which allows high level synthesis of the *cI* repressor while the *cIII*-gene product protects the *cII*-protein from the proteolytic activity of a bacterial protease coded by the *hfl* locus.

M13 and Mu1: Phage Important in Recombinant DNA Techniques

M13 is a single-stranded DNA filamentous phage related to fd, and Mu is a strange phage which combines the properties of a temperate phage and a transposable genetic element. M13 is useful because it can be used to clone, amplify and separate single strands of DNA which can then be used for heteroduplex analysis or for base sequencing. There is also a double-stranded replicative intermediate which can be handled with the same techniques as for other DNA molecules. Mu phage has been used to promote mutation by insertion, to produce deletions and inversion, and also to produce fusions between *replicons* (units for replication). M13 is unusual in its mode of infection and release. Protein coded by gene-3 are required for adsorption and release, both of which occur by the filamentous phage passing through wall and the membrane without lysing the cell. The gene 5-protein protects the single-stranded DNA prior to its packaging into the coat protein in the membrane of the cell. M13 produce plaques as infection shows the growth of the host cells giving thinner areas of bacterial lawn.

11

Plasmids

Plasmids are extrachromosomal circular DNA molecules found in most bacterial species and in some species of eukaryotes. Under normal circumstances a particular plasmid is dispensable to its host cell; for example, sometimes at the time of cell division a plasmid-free daughter cell is formed and such a cell is almost always viable. However, many plasmids contain plasmid genes that may be essential in certain environments. For example, the R plasmids carry genes that confer resistance to numerous antibiotics so that in nature a cell containing such a plasmid can survive in the presence of an antibiotic, whether humanly administered or produced by a fungus.

Transposable elements (or transposons, as they are called in bacteria) are another type of accessory DNA molecule. They are chromosomal segments that differ from other regions of the chromosome in that they are able to relocate (transpose) to another part of the chromosome or to a plasmid, virus, or separate chromosome. Their movement, which occurs infrequently and usually at random in time, is mediated by enzymes encoded in the element; other enzymes, for instance, some conferring drug resistance, are contained in some elements.

Currently, major interest in plasmids centers on their practical value in genetic engineering. Transposable elements are responsible for many genetic phenomena in both bacteria and eukaryotes; in the latter, they are involved in the regulation of the activity of some genes.

Plasmid DNA

With only a single exception (the killer-plasmid of yeast, which is an RNA molecule) all known plasmids are supercoiled circular DNA

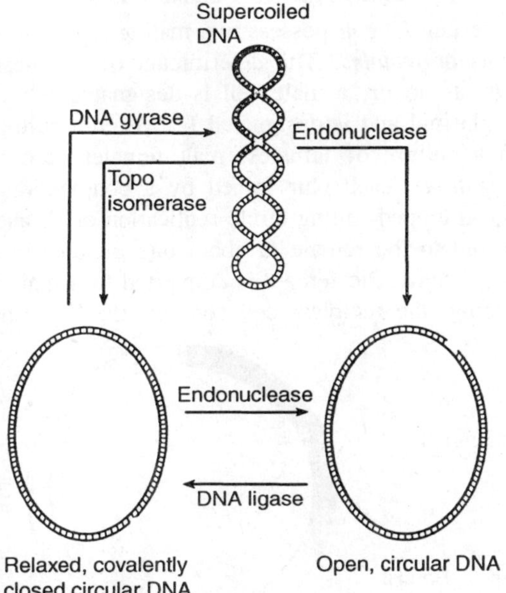

Fig. 11.1. The interconversion of supercoiled, relaxed covalently closed circular DNA and open circular DNA.

molecule. The molecular weights of the DNA range from about 10^6 for the smallest plasmid to slightly more than 10^8 for the largest one. The number of copies of a plasmid per cell ranges from 1–2 (low copy number plasmids) to 10–60 (high-copy-number plasmids).

Plasmid DNA can usually be isolated from bacteria in a simple way. A culture of plasmid-containing bacteria is lysed by adding a detergent and then the lysate is centrifuged. The bacterial chromosome complex, which contains protein and RNA, is very large and compact and moves rapidly to the bottom of the centrifuge tube; the smaller plasmid DNA remains in the supernatant. CsCl and the fluorescent dye ethidium bromide are then added to the supernatant and the solution is centrifuged to equilibrium. The CsCl forms a density gradient. In the presence of ethidium bromide supercoiled molecules have a higher density than the linear fragments of chromosomal DNA (which is broken during isolation) ; thus, at equilibrium the supercoiled DNA is located in a different region of the centrifuge tube than the linear molecules. The fluorescence of the ethidium bromide makes both DNA fractions visible, so the supercoils can be easily identified and removed from the tube.

Transfer of Plasmid DNA

The bacterium *E. coli* possess two mating types: donors or *males*, and recipients or *females*. The determinant of maleness is the F or *sex plasmids*; as donor, a male cell is designated F^+. A female cell lacks the F plasmid and is designated F^-. When a culture of males is mixed with a culture of females, male-females pairs form (this is called *conjugation*), each pair joined by a conjugation bridge. This pairing induced looped rolling circle replication of F, and one copy of F is transferred to the female in about one minute. In contrast with other sexual systems, the female is converted to a male inasmuch as, after the mating, the recipient cell contains the F plasmid.

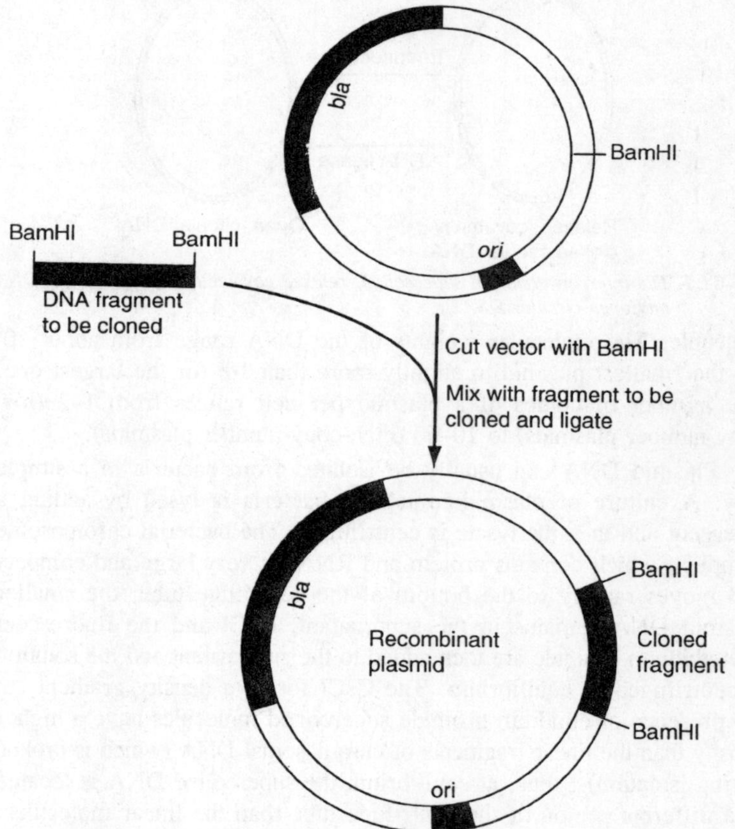

Fig. 11.2. Cloning with a plasmid vector: bla= beta-lactamase (ampicillin resistance) selective marker; ori = origin of replication.

Note that during transfer DNA synthesis occurs in both the donor and recipient cells. In the donor the synthesis replaced the single strand that is transferred; synthesis in the recipient converts the transferred single strand to double-stranded DNA. The mechanism of recircularization of the transferred DNA is not well understood. It is believed that when one cycle of looped rolling circle replication is completed, the protein, that binds to the 5' terminus makes a cut at the replication origin and brings the 3' and 5' termini together in preparation for a joining event.

Many, but not all, classes of plasmids are capable of transfer. Plasmids unable to transfer usually either are unable to form stable pairs or lack a system for transfer replication.

Another property possessed by the F plasmid is the ability to integrate into the bacterial chromosome. When this occurs, the chromosome remains a single, circular DNA molecule and F behaves as if it were part of this chromosome, increasing the size of the chromosome. The integration process happens rarely but it is possible to purify a culture of cells that are the progeny of the cell in which integration occurred. The cells in such a culture are called *Hft males*. (Hfr is an acronym for high frequency of recombination.) When a culture of Hfr cells is subsequently mixed with an F^- culture, conjugation also occurs as described earlier, though the material transferred is slightly different from that in an F^+ x F^- mating. This time, under the influence of F, DNA replication begins in the Hfr cell and a replica is transferred to the F^- cell; however the direction of replication is such that a small portion of F is transferred first and the major portion is transferred last. Moreover, because bacteria are very small and in constant motion by being bombarded by solvent molecules (Brownian motion), and because it takes 100 minutes to transfer an entire chromosome, the mating pair usually breaks apart before transfer is completed. Thus, the female receives both a large fragment of the male chromosome, which may contain hundreds or thousands of genes, and a small functionless fragment of F. Because of this the exconjugant female remains a female in an Hfr x F^- mating.

The presence of the new chromosomal fragment in the female sets in motion a recombination system that causes genetic exchanges to occur, so that a recombinant F^- cell often results. Thus, in a mating between and Hfr leu^+ culture and an F^-leu^- culture F^-leu^+ cells form. Hfr x F^- matings are powerful matings for creating bacteria with desired combinations of mutations.

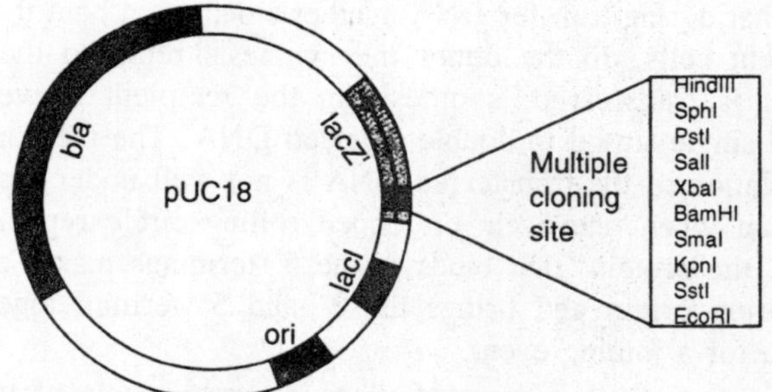

bal = beta-lactamase (ampicillin resistance); selective marker
ori = origin of replication
lacZ' = beta-galactosidase (partial gene)
lacI = repressor of lac promoter

Fig. 11.3. Structure of the plasmid cloning vector pUC18.

An Hfr is produced when F integrates stably into the chromosome, as we have already stated. At a low frequency, F can excise itself. When this happens, the excised circular DNA is sometimes found to contain genes that were adjacent to F in the chromosome. A plasmid containing both F genes and chromosomal genes is called F` plasmid. It is usual to describe an F` plasmid by stating the genes it is known to possess–for example, F' *lac pro* contains the genes for lactose utilization and proline synthesis. F' plasmids can also be transferred from an F' male to a female. This occurs sufficiently rapidly that the entire F' is usually transferred before the mating pair breaks apart. Thus the female recipient is converted to an F' male in an F' x F^- mating.

F' plasmids have been useful in the production of partial diploid bacteria. For example, if an F'*lac*$^+$/*str-s* male is mated with *lac*$^-$*str-r* female and a Lac+Str-r colony is isolated the cells in the colony will carry two copies of the *lac* gene—the *lac*$^+$ version brought to the female in the F' and the *lac*$^-$ version already present in the female chromosome. To denote this, the genotype of the cell is written F'*lac*$^-$ /*lac*$^-$ *str-r*, as is usual with partial diploid cells. By convention, genes carried on the F' plasmid are written at the left of the diagonal line. F' plasmids have been exceedingly useful in the analysis of regulation of bacterial system.

Often it is desirable to transfer a plasmid that is incapable of conjugal transfer, for example, the Col plasmids. When this is required, the plasmid can be transferred, albeit at much lower efficiency, simply by mixing free plasmid DNA with recipient cells suspended in a cold

$CaCl_2$ solution, a procedure known as *$CaCl_2$ transformation*. The main requirement for successful transfer is that the recipient bacterium possess replication enzymes that are active on the plasmid, which is invariably the case if the donor and recipients are of the same species; sometimes, interspecific transformation is also successful.

Properties of Particular Plasmids

We have just seen some of the properties of the F plasmid. In this section we briefly describe some of the more commonly studied plasmids. These have been used primarily as means of studying DNA replication and as a cloning vehicle in genetic engineering.

The drug-resistance, or R, plasmids were first isolated from the bacterium *Shigella dysenteriae* during an outbreak of dysentery in Japan and have since been found in *E. coli* and various species of *Salmonella*, *Vibrio*, *Bacillus*, *Pseudomonas*, and *Staphylococcus*. Their defining characteristics are that they confer resistance on their host cell to a variety of fungal antibiotics and are usually self-transmissible. Most R plasmids consist of two contiguous segments of DNA. One of these segments is called *RTF* (*resistance transfer factor*); it carries genes regulating DNA replication and copy number, the transfer genes, and sometimes the gene for tetracycline resistance (*tet*), and has a molecular weight of 11×10^6. The other segment, sometimes called the *r determinant*, is variable in, size from a few million to more than 100×10^6 molecular weight units) and carries other genes for antibiotic resistance. Resistance to the drugs penicillin (Pen), ampicillin (Amp), chloramphenicol (Cam), streptomycin (Str), kanamycin (Kan), and sulfonamide (Sul), in combinations of one or more, appears commonly.

Col plasmids are *E. coli* plasmids encoding colicins, proteins that are capable of preventing growth of a bacterial strain that does not contain a Col plasmid. There are many types of colicins, each designated by a letter (e.g., colicin B) and each having a particular mode of inhibition of sensitive cells. The best-studied Col plasmid is Col El, whose molecular weight is 4.2×10^6. It is used extensively in recombinant DNA research and in an in vitro DNA replication system.

An crown gall tumor found in many dicotyledonous plants is caused by the bacterium *Agrobacterium tumefaciens*. The tumor-causing ability resides in a plasmid called *Ti*. In an infected plant some of the bacteria enter and grow within the plant cells and lyse there, releasing their DNA in the cell, and from this point on, the bacteria are no longer necessary for tumor formation. By an unknown mechanism a small fragment of the Ti plasmid, containing the genes for replication, becomes

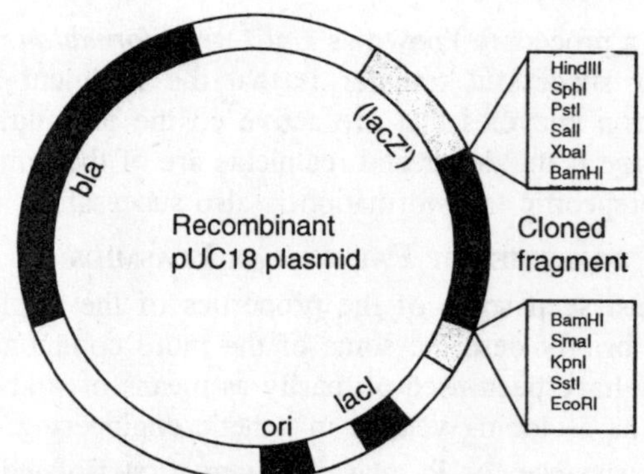

The lacZ gene has been disrupted by insertion of a DNA fragment, resulting in white colonies on X-gal plates

bla = beta-lactamase (ampicillin resistance); selective marker
ori = origin of replication
lacZ' = beta-galactosidase (partial gene)
lacI = repressor of lac promoter

Fig. 11.4. Use of the plasmid cloning vector pUC18.

integrated into the plant cell chromosomes. The integrated fragment breaks down the hormonally regulated system that controls cell division and the cell is thereby converted to a tumor cell. This plasmid has recently become very important in plant breeding because specific genes can be inserted into the Ti plasmid by recombinant DNA techniques; sometimes these genes can become integrated into the plant chromosome, thereby permanently changing the genotype and phenotype of the plant. It is believed that new plant varieties having desirable and economically valuable characteristics derived from unrelated species can be developed in this way.

Plasmids are known in yeast, unicellular eukaryote. One of the more intriguing ones is the *killer particle*, which is a double stranded RNA molecule of molecular weight about 1.5×10^6. It is the only known plasmid that does not contain DNA. This particle contains ten genes for replication and several others for synthesis of the killer substance, a colicin-like material.

The best-studied yeast plasmid is the so-called 2 μm plasmid, a DNA molecule having a length of 2 μm and a molecular weight of 4×10^6. There are about sixty copies of this plasmid per cell and all are found in the yeast nucleus. This location is consistent with the fact that for replication the 2 μm plasmid uses the same enzymes needed for replication of the yeast chromosomes. Furthermore, the plasmid DNA is coated with histones, as is all nuclear DNA. The plasmid DNA apparently does not integrate into the host DNA.

Plasmid Replication

A plasmid can only replicate within a host cell; one might therefore expect that all plasmids native to the same host species would have the same mode of replication. However, just as with phages, which also replicate only within a host cell, there is enormous variation in both the enzymology and mechanics of plasmid DNA replication. The major types of variability are the following:

1. *Reliance on host enzymes*. Some plasmids use host enzymes exclusively. Others encode some of their enzymes.
2. *Identity of polymerase*. Most *E. coli* plasmids use pol-III for chain growth and pol-I for synthesis of precursor fragments, as in replication of the E. coli chromosome. However, others (Col E1 is and example use pol-I for chain growth.
3. *Directionality*. Both purely unidirectional and purely bidirectional replication have been observed. In addition, there are plasmids in which both modes are present. For example, the plasmid RK6 replicates first in one direction and then later in the opposite direction from the same origin.
4. *Termination*. In unidirectionally replicating plasmids termination necessarily occurs at the origin after one cycle of replication. Bidirectionally replicating molecules are of two types though, both of which have been observed. In one type, termination occurs when the growing forks reach the same region. Others have a fixed termination site that is sometimes reached by one growing fork before the other fork reaches it. The signal for termination is unknown.
5. *Replicating form*. In the most carefully studied plasmids, replication occurs by the so-called *butterfly* mode, first observed for animal viruses. In a partially replicated molecule the replicated portions are untwisted, as is usually the case in θ replication, but the unreplicated portion is supercoiled. When the replication cycle is completed, one of the circles must be cleaved (possibly by DNA gyrase). The result after one round of replication is one nicked molecule and one supercoiled molecule. The nicked molecule is then sealed and, somewhat later, supercoiled. Whether this is a general mechanism for plasmid replication is not known.

Control of Copy Number

Some plasmids are present in cells in low-copy number—one or a few per cell—whereas other exist in large numbers–from 10 to 100 per cell. The number is regulated by controlling the rate of initiation of DNA synthesis.

The generally accepted explanation for the regulation of copy number is that there is a plasmid-encoded inhibitor that binds to the replication origin and inhibits replication. Let us first see how this idea explains maintenance of copy number from generation to generation. In the current theory it is assumed that the inhibitor is active only as a multisubunit protein and that the monomer-multimer equilibrium is very dependent on concentration. Thus, as a cell grows (enlarges), the inhibitor concentration drops and inactive monomers form ; replication is thereby depressed, and the number of plasmid DNA molecules doubles. At this point there will exist twice the initial number of inhibitor genes; therefore, by protein synthesis the inhibitor concentration also doubles. This results in the formation of active inhibitor and replication stops. A similar sequence of events would occur if there were initially only one copy of a high-copy-number plasmid–that is, replication would continue until there is sufficient inhibitor to turn off synthesis. How can this explanation account for differences in copy number? The most likely possibility is that for the high-copy-number plasmids the association of monomers to form a multimeric active inhibitor requires a higher concentration of monomers than is the case for the low-copy-number plasmids. Thus, only when the number of plasmids per cell is high is the "gene dosage" high enough to produce active inhibitor.

Structure of Transposable Elements

Transposable elements were first detected in *E. coli* as polar mutations in the *gal* operon. In order to understand these mutations, fragments of DNA containing such mutations were isolated and their nucleotide sequences were determined. It was found that each polar mutation was associated with a segment of DNA that had somehow been inserted in the gene. These segments were called *insertion IS elements* and later shown to be members of a class of sequences called *transposable elements* or, in bacteria, *transposons*.

A variety of different IS elements, ranging in length from a few hundred to a few thousand nucleotide pairs, have been isolated and sequenced. A feature of all IS elements is the presence of inverted-repeat sequences at the termini. The polar effects are the result of either transcription termination signals or stop codons (in all reading frames) in the main body of the element. IS elements are found in various regions of the *E. coli* chromosome, in plasmids, and in some phages. They have also been found at different sites within a single gene, inserted in both orientations : having both orientations is not unexpected in view of the symmetry of the termini. The feature that has aroused greatest interest in IS elements is their ability to move from one site to another, which is the origin of the term transposable element. This movement, called *transposition*, is the defining characteristic of transposons and will be described shortly.

Table 11.1. Properties of some *E. coli* insertion elements.

Elements	*Number of copies and location*	*Size, base pairs*
IS1	5-8 in chromosome	768
IS2	5 in chromosome, 1 in F	1327
IS3	5 in chromosome, 2 in F	Approximately 1400
IS4	1 or 2 in chromosome	Approximately 1400
IS5	Unknown	1250
$\gamma\delta$	1 or more in chromosome, 1 in F	5700

Since the discovery of the IS elements numerous transposons carrying genes that are easily identifiable–in particular, antibiotic-resistance genes—have been found in several bacterial species. These transposons have been studied extensively since their presence can be ascertained merely by noting whether the host cell can form a colony on growth medium containing a particular antibiotic. The transposons containing accessory genes are of two sorts—the composite and the Tn3 families.

The composite family

Transposons in this family consist of two identical IS elements that flank the antibiotic resistance segment. The IS elements in these composite units can be in a inverted or direct-repeat configuration. Note that since the two ends of an IS element are themselves inverted repeats, the relative orientation (inverted or direct) of the flanking IS elements of a composite transposon does not alter its terminal sequences –they remain the same in the inverted-repeat array.

The Tn3 transposon family

The Tn3 family of transposons consists of quite large elements (about 5000 base pairs). Each transposon carries there genes, one encoding β-lactamase (which confers resistance to ampicillin) and two others needed for transposition (which will be discussed later). All Tn3-like transposons contain short (38-base-pair) inverted repeats and none are flanked by IS-like elements.

Transposition

Consider a culture of a particular bacterium that contains a single transposon at some site in the chromosome. If this culture is grown for many generations, rare mutant bacteria will result in which the transposon is present within genes at sites different from the initial site. This apparent movement exemplifies the phenomenon of transposition, "apparent" because genetic and physical analysis usually show that such mutant cells contain two copies of the transposon—one at the original site and another at the new site (this will be discussed shortly). On continued growth of a cell containing two copies, another rare transposition event may occur, leading to a cell with three copies. Ultimately, all cells would contain an enormous number of copies of a particular transposon were it not for the fact that transposons occasionally are excised and lost.

Transposition is usually not detected by mutant formation, for technical reasons and because insertion does not always occur within genes. Instead, one normally works with transposons carrying antibiotic-resistance markers and looks for their apparent movement from one DNA molecule to another for example, from the chromosome to a plasmid or form a plasmid to a phage. Since transposition is a recombination event, bacterial mutants defective is normal genetic recombination are used—for example, bacterial *recA*⁻ mutants; in this way, one can study movement that is controlled only by the transposon itself—namely, transposition.

An R plasmid containing an *amp-r* transposon is transferred by conjugation to a *recA*⁻ recipient cell that carries another transposon having the *kan-r* (kanamy-cin-resistance) gene. On continued growth of a culture containing donors, recipients, and conjugated cells in the presence of both antibiotics (so that neither donor nor recipient can multiply), only cells that have received the R plasmid can grow. In these cells transposition of the *kan-r* transposon from the recipient chromosome to the R factor occasionally occurs, yielding an R plasmid carrying both antibiotic-resistance markers. One can demonstrate that the kan-r transposon is carried by the R factor by mating these cells with a Kan-s Amp-s culture (which carries another antibiotic-resistance marker so that donor and recipient cells can be distinguished), allowing conjugation to occur, and plating on a medium containing all there antibiotics. Only recipient cells that have obtained both the *kan-r* and *amp-r* markers together on the R plasmid can form colonies.

When transposition occurs, a particular transposon can usually the inserted at one of a large number of positions. For example, if an *amp-r* transposon is transferred to *lac*$^+$ *leu*$^+$ *amp-s* cells, in time both *lac*$^-$ *leu*$^+$ *amp-r* and *lac*$^+$ *leu*$^-$ *amp-r* cells will accumulate in the culture. However, the insertion sites are not randomly distributed at the nucleotide level, and different transposons have various degrees of selectivity for particular sites. At one extreme is IS4, for which 20 independent insertion events have been observed in the *E. coli galT* gene at a single site. In transposition of the element Tn9 into the terminal 160 bases of the *E. coli* lacZ gene, 28 independent transpositions were detected; 16 positions were observed but 5 of these were represented by multiple occurrences of insertion. The least selective element is Mu, which apparently can insert at any site. The cause of selectivity in some cases is not known.

Mechanism of Transposition

The end of the transposition process is the insertion of a transposon between two base pairs in a recipient DNA molecule. Base sequence analysis of many transposons and their insertion sites reveals that there is no sequence homology, which is consistent with the lack of a requirement for the *E. coli recA* system. However, the sequences of the regions in which the inserted element joins the recipient DNA has yielded several surprises.

A characteristic of the insertion process is that insertion of a transposon always involves the duplication of a short base sequence (3–12 base pairs long) in the recipient DNA molecule, called the *target sequence*, and the inserted transposon is sandwiched between the repeated bases. We repeat, emphatically, that *only one copy of the duplicated targed sequence is present in the recipient DNA prior to insertion of the transposon and it is not present in the transposon itself.* The length of the target sequence varies from one transposon to the next but *is the same for all insertions of a particular transposon.* For example, an inserted ISl is always flanked by a nine-base-pair sequence, whereas Tn3 is always flanked by a duplication of five base pairs. Note that *for a particular transposon, the largest sequence is different for each insertion site; only the length of the duplicated sequence is constant.*

A mentioned earlier, in bacterial transposition one copy of the transposon remains at the original site. In which transposition between plasmids is observed. After many generations of growth of a bacterium containing both plasmid A, which has a transposon, and plasmid B, which lacks a transposon, cells are produced in which plasmid B (now called B') also possesses the transposon. This observation is significant in understanding the mechanism of transposition, because it indicates that the original transposon is duplicated in the transposition process. Thus, both the target sequence and the transposon are duplicated, which means that transposition is a *replicative process*.

Some insight into transposition is obtained from the following phenomenon. Consider a cell with two plasmids, one of which contains a transposon. At a frequency of about 10^{-7} events per cell generation, the two plasmids fuse to form a single plasmid called a cointegrate (this process is sometimes also called *replicon fusion*.) This hydrid plasmid is not simply a fusion of the two plasmids, because it contains two (not just one) copies of the transposon. Both copies are in the same orientation (that is, in direct repeat) and precisely at the junction between the donor and recipient plasmid sequences. As far as is known, all transposons are capable of mediating cointegration though the cointegrates are not always stable.

The formation of cointegrates is significant for two reasons. First cointegrates are thought to be an essential intermediate in the transposition process and, second, the fact that the cointegrate has two copies of the transposon, whereas only one copy was present before cointegration occurred, again indicates that DNA replication of the transposon has occurred.

The cointegrate as an intermediate is suggested by certain features of Tn3 having two inverted repeats, like all transposons, and these flank three genes. The three gene products have been isolated and from their size and the base sequence of Tn3, we know that the genes utilize all of the bases between the inverted repeats. The leftmost gene encodes a gigantic protein (1015 amino acids), denoted TnpA, which is a transposase and which is responsible for formation of cointegrates. The rightmost gene encodes b-lactamase, an enzyme that inactivates ampicillin. The central gene encodes a small protein (185 amino acids) called TnpR, which has two functions—negative regulation or repression of the synthesis of TnpA, and promotion of a site-specific exchange that resolves cointegration in a second step of the transposition process. Near the boundary of the genes encoding TnpA and TnpR is a sequence of DNA called the *internal resolution site* and it includes the base sequence at which TnpR acts.

Genetic experiments have yielded the following information. All tnpA—mutants are unable either to transpose or to form cointegrates. Also, *tnpR*⁻ mutants and deletions of the internal resolution site form cointegrates yet are unable to transpose. That is, in a cell containing a plasmid with *tnpR*⁻ copy of Tn3 and a second plasmid lacking Tn3, the process shown can occur, but the process shown does not occur. These observations suggest that Tn3-mediated transposition occurs by a two-step process—first, TnpA induces formation of a cointegrate and then TnpR promotes a site-specific exchange at the internal resolution site. Some transposons do not carry a tnpR-like site specific recombination system but are still able to transpose. These systems presumably use a homology-dependent exchange system since, once a cointegrate has formed, there are two identical sequences—namely, copies of the transposon—so that any homology-dependent process could carry out the exchange.

Transposable Elements in Eukaryotes

The first evidence for the existence of transposable elements came not from bacteria but from ingenious genetic studies of maize by Barbara McClintock. Her observations indicated that the activities of certain genetic loci in corn are distributed by a controlling element that inhibits the activity only when the element is adjacent to the locus change in level of gene activity seemed to occur by movement of the element to and from its inhibiting position. The controlling element, called *dissociation*, has been found to be a transposable element, and its movement is regulated by a second transposable element called *activator*. Both of these have been isolated and their base sequences determined.

Transposable elements have been observed in other eukaryotes. The ye yeasts *Saccharomyces cerevisiae* contains several different transposable elements of which the best-studied are the IS-like element δ and the composite element Ty1 (for which about 35 copies are present per haploid cell). In the fruit fly *Drosophila melanogaster* 5–10 percent of the DNA consists of 30–40 distinct families of transposable elements. The best understood ones are copia and the P elements. Transposition of copia affects the activity of certain genes, presumably by movement of regulatory sites, such as promoters and other protein-binding sites.

A number of genetic results with *Drosophila* can be explained by the existence of transposons. For example, there are mutations that inhibit nearby recombination events, are polar, and have been shown to be insertions.

An important possibility is that some (or perhaps all) of the RNA tumor viruses are themselves transposons : however, the only evidence for this view at present is that several integrated tumor viruses are flanked by 600-base pair sequences in direct repetition and, in addition, a duplication of a short target sequence.

Genetic Phenomena Mediated by Transposons

Transposable elements mediate a variety of genetic phenomena, such as gene rearrangements (for example, inversions), deletions and fusion of DNA molecules (as in cointegrate formation). Transposable elements can also serve as switches turning genes on and off (for example, *Dissociation* in maize) or by inverting sequences containing promoters. These phenomena are direct consequences of transposition, and some have already been discussed; hypotheses for the mechanisms for producing inversion and deletions can be found in MB.

Several processes in bacteria are known to be the result of physical exchange (recombination) between multiple copies of transposons. Two of these, formation of Hfr cells and gene amplification, are described in this section.

Hfr cell are formed by integration of F, as has already been stated. Critical to understanding how this occurs are two facts: (1) F contains several IS elements and (2) in an Hfr an integrated F is always flanked by two copies, in direct respect, of one of the IS elements of F. The primary mechanism for formation of Hfr cells is a reciprocal exchange between an IS element in F and the same element in the chromosome. Since the IS elements are found at various sites in the chromosome and since they are capable of relocating by transposition, numerous integration sites are possible. Furthermore, since transposed IS elements in the chromosome can be in two different orientations, F can be integrated both clockwise and counter clockwise with respect of the genetic map of *E. coli*. This explanation is sufficient to understand the origin of the many different Hfr cell lines and the fact that Hfr strains that transfer DNA in both directions with respect to the *E. coli* map are known.

Another phenomenon in which homologous recombination between multiple copies of transposons plays a role is gene amplification in bacteria. Some R plasmids have the property that if a bacterium containing the plasmid is exposed to an antibiotic concentration considerably lower than the maximum concentration tolerated by the cell, over a period of many generations the cell becomes resistant to ever-increasing concentrations of the antibiotic. This result from a gradually increasing number of antibiotic-resistance segments in the R plasmid. That is, the R plasmid increases in size owing to repeated tandem duplications of the antibiotic resistance genes. This process requires an active bacterial Rec system, showing that duplication is not a transposition process. Based on the fact that the antibiotic-resistance segment or R plasmids are flanked by transposons, two models for gene amplification have been developed. None that the two basic requirements of the models are that the genes to be amplified are flanked by identical sequences and that the Rec system acts effectively on pairs of homologous regions.

INDEX